The cradle of the mind, but we cannot live forever in a cradle

A Firefly Book

Published by Firefly Books Ltd. 2004

First printing

Publisher Cataloging-in-Publication Data (U.S.)

Chaikin, Andrew.
    Space : a history of space exploration in photographs / Andrew Chaikin; foreword by Captain James A. Lovell.
Originally published: London: Carlton Books, 2002.
[256] p. : photos. (chiefly col.) ;   cm.
Includes index.
Summary: Detailed account of the developments that have taken place in space exploration over the past six decades.
ISBN 1-55297-987-3 (pbk.)
1. Astronomy. 2. Outer space – Exploration.   I. Lovell, James A. II. Title.
520 22   QB44.3.C53 2004

National Library of Canada Cataloguing in Publication

Chaikin, Andrew, 1956-
        Space : a history of space exploration in photographs / Andrew Chaikin ;
foreword by James A. Lovell.
Reprint. Originally published: London : Carlton, 2002.
ISBN 1-55297-987-3
        1. Outer space—Exploration—Pictorial works. I. Title.
QB500.262.C43 2004   629.4'1   C2004-902157-5

Published in the United States in 2004 by
Firefly Books (U.S.) Inc.
P.O. Box 1338, Ellicott Station
Buffalo, New York 14205

Published in Canada in 2004 by
Firefly Books Ltd.
66 Leek Crescent
Richmond Hill, Ontario L4B 1H1

Cover photographs courtesy of NASA

Printed in China

Page 1. Self-portrait of a lunar pioneer: *Surveyor 1*, which soft-landed on the Moon in June 1966, and televised this view of its own shadow stretching across the dusty lunar surface.

Page 4. Heading for the International Space Station, the space shuttle *Atlantis* thunders off the launch pad as the STS-101 mission begins with a pre-dawn liftoff on May 19, 2000.

Page 5. Cosmonaut Leonid Kizim adjusts his sun visor during one of nine spacewalks he made as part of his eight-month stay aboard the *Salyut 7* space station in 1984.

Page 6. Saturn's largest moon, Titan, is the only satellite in the solar system known to possess a dense atmosphere, which encircles the backlit moon in this image taken by *Voyager 2* on August 25, 1981.

Pages 10–11. The spectacle of an orbital sunset, captured in a telephoto view from the space shuttle *Endeavour* in September 1992.

Page 13. Nearly four decades after Ed White became the first American to walk in space, Pat Forrester works in the cargo bay of the space shuttle *Discovery* during the STS-105 mission to the International Space Station, August 16, 2001.

SPACE

A History of Space Exploration in Photographs

Andrew Chaikin

Foreword by Captain James A. Lovell

FIREFLY BOOKS

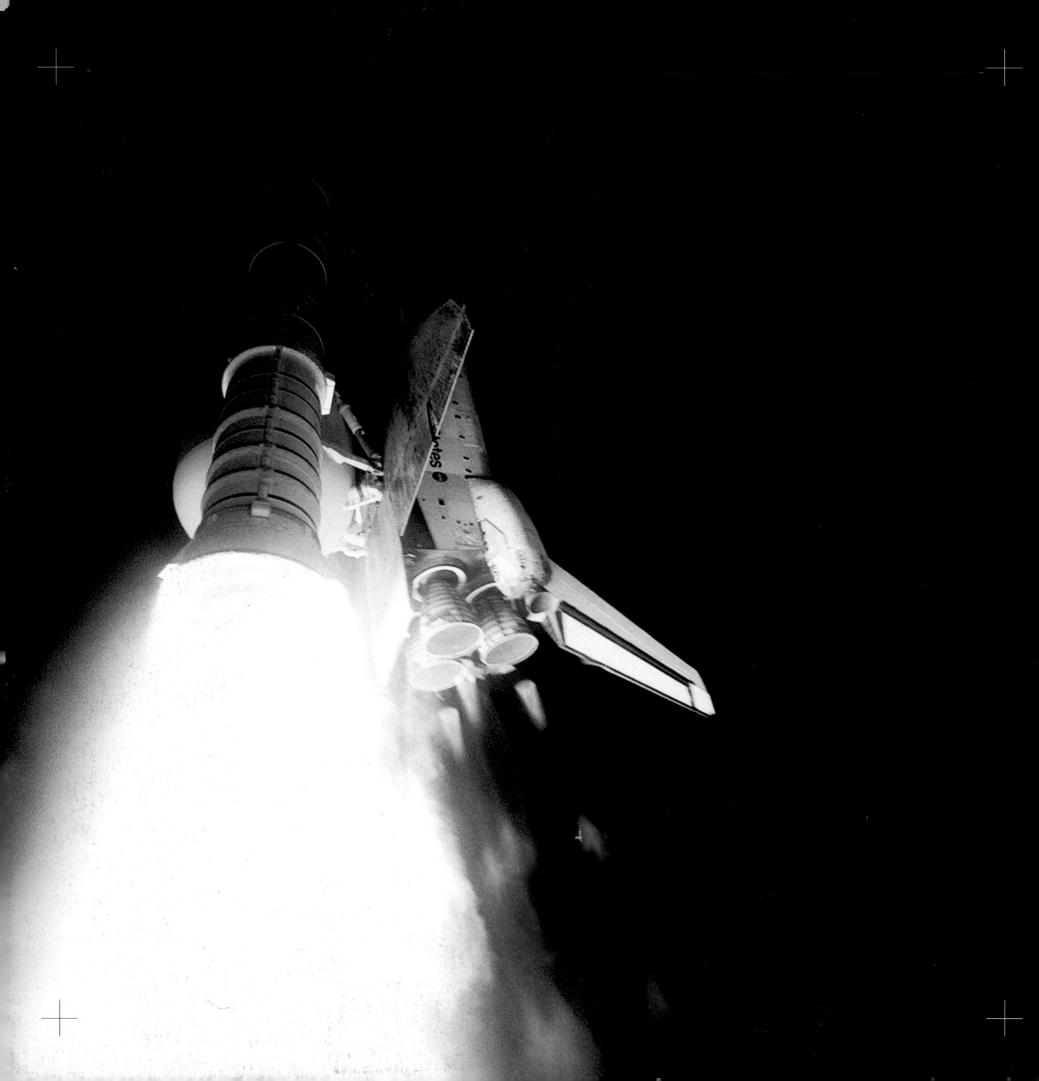

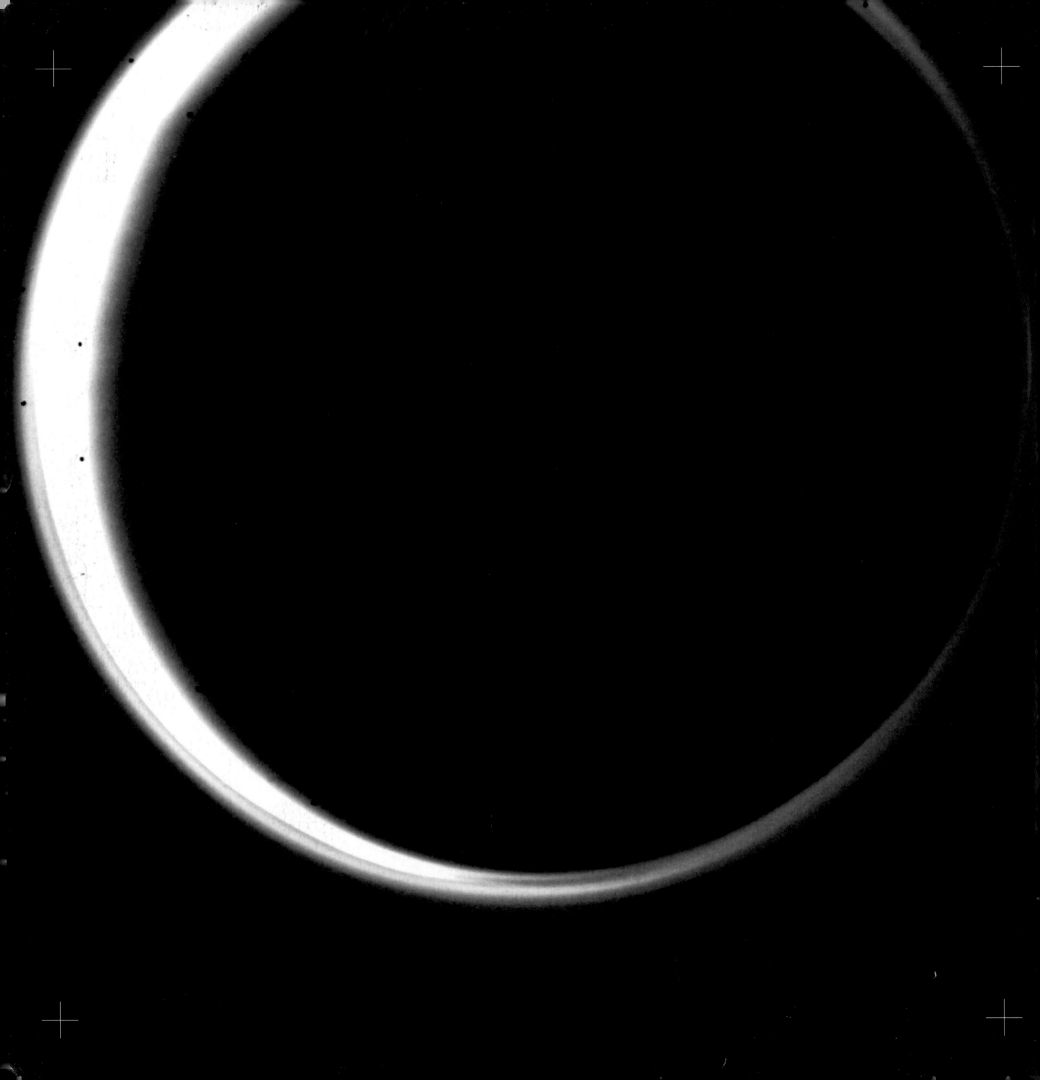

FOREWORD BY
CAPTAIN JAMES A. LOVELL

Andy Chaikin, who brought us the incredible adventures of the Apollo Moon missions in his previous book *A Man on the Moon*, now widens the scope of our knowledge to the world's conquest of space with this excellent book. He takes us from the origins of spaceflight, where we meet the early pioneers, Tsiolkovsky, Goddard and Korolëv and see their triumphs and tragedies. We learn about the human flights of the Mercury, Gemini, Vostok and Soyuz programmes; the unmanned probes to Venus and Mars; the success of Apollo and the failure of the Russian Lunar Programme. The text and photos rejuvenate the excitement first felt when *Sputnik* surprised the world; when Yuri Gagarian made the first Earth orbit and when Neil Armstrong landed on the Moon.

Picture yourself in Armstrong's shoes, as he surveys the lunar landscape just after landing. The view appears uninviting, forbidden, hostile and he wonders: "What kind of day will it be?"

It will be a day like all days on the surface of the Moon – a velvet black sky with a terrain of rocks, boulders, craters and dust in all shades of gray. The only colour to be seen – the Earth – so distant and small it can be hidden behind his thumb. A vacuum atmosphere surrounds him, as he "bunny hops" from place to place in the one-sixth "g" gravitational environment. Moon dust, kicked up by his footsteps, falls back to the surface with mathematical precision without an atmosphere to disturb it. It is a silent world. Only the sound of his life support system, the crackling of communications and the echo of his breathing in the fish-bowl helmet keep him company.

This day Armstrong will remember forever and through these pages you too will relive the excitement of his first steps on the Moon and many of the other incredible achievements of the space age.

So, fasten your seatbelt and launch into space with the first page. You will be mesmerized by Andy Chaikin's informative text and the remarkable photos as you journey through this book until the last page returns you safely back to Earth.

Captain James A. Lovell
Gemini 7 & 12, Apollo 8 & 13

AUTHOR'S PREFACE

My life changed at the age of nine in June 1965, when the new issue of *Life* magazine arrived with some of the most spectacular photographs I'd ever seen. They were pictures of *Gemini 4* astronaut Ed White emerging into the vacuum of space. He was in brilliant sunlight, and yet the sky was pitch black. Beyond him was a vivid blue, curved horizon: the edge of the Earth. Looking at him floating above the planet, I felt a burning desire to follow in those weightless footsteps. I was hooked. I wasn't the only one who felt the power of those images from space. Countless people saw them and understood their basic message: this was the edge of human experience.

And that edge kept moving. Each new mission helped push humanity farther from home, culminating in the first voyages to the Moon by the Apollo astronauts. These historic explorations also produced some of the twentieth century's most incredible photographs, including *Apollo 8*'s snapshot of the Earth rising above the Moon's battered and lifeless horizon on Christmas Eve, 1968. Months later, in July 1969, the dream of centuries came true, and we saw the first photographs taken by men on the Moon.

At the same time, space probes began sending back pictures from places where humans could not go, including every planet in our solar system except Pluto, a menagerie of moons and even a handful of comets and asteroids. Worlds that were just specks of light in telescopes have become places we can name and study, with fantastic landscapes. And the Hubble Space Telescope has allowed us to see even farther, to the very edge of the universe. Again and again, these amazing machines have given us the thrill of seeing things no one has ever seen before.

This collection of photographs is an effort to share that thrill and tell the story of space exploration, through images that are not just historically significant but also beautiful. Individually, these pictures have the power to take us outside our everyday lives. Collectively, they tell a remarkable story of human curiosity, ingenuity and persistence – qualities that have enabled us to expand our boundaries beyond our home world. As you turn the pages of this book, think about those qualities, and ask yourself if they don't make you feel lucky to be alive at such an extraordinary time.

Andrew Chaikin

INTRODUCTION
DREAMS OF SPACE

We live in an age when space exploration is an everyday occurrence. Men and women are living and working in space for months at a time aboard an International Space Station that circles the Earth some 250 miles (400 kilometres) up. Space probes send back stunning images of distant worlds in our own solar system, while in Earth orbit the Hubble Space Telescope peers even farther out into space, probing the mysteries of the cosmos. At times it seems as if it has always been this way.

The reality is quite different. It wasn't so long ago – the mid-1950s – that the very idea of sending humans into space seemed so fantastic that most people doubted it would happen in their lifetimes. No one could have guessed that Soviet cosmonaut Yuri Gagarin would become the first human to orbit the Earth in 1961, or that American astronauts Neil Armstrong and Edwin "Buzz" Aldrin would leave mankind's first footprints on the Moon just eight years later. Even in retrospect, the advent of the space age is among the most extraordinary events in human history. But while its causes were largely unforeseen, it didn't happen by accident. Its origins lie not only in the insights of some of history's greatest scientists and the efforts of the twentieth-century's space pioneers, but in the geopolitics of the Cold War.

Nobody knows when humans first dreamed of exploring the realms beyond Earth, but there are stories of space travel going back as far as the second century A.D., when Lucian of Samosata, a Greek satirist, described a sailing ship lifted from the ocean and carried to the Moon by a violent whirlwind. Some 1,400 years later, in 1649, the celebrated French philosopher and playwright Cyrano de Bergerac penned his own fanciful lunar journey in *Voyage dans la Lune* (*Voyage to the Moon*) followed by a sequel, *Histoire Comique*

des Etats et Empires du Soleil (*Comic History of States and Empires of the Sun*). De Bergerac, while acquainted with the scientific theories of the day, let his imagination run wild.

However, it was a Russian mathematics teacher named Konstantin Tsiolkovsky who laid the groundwork for space travel. Largely self-educated, Tsiolkovsky had been profoundly deaf from childhood; in his brilliant writings and forecasts, written between 1892 and his death in 1935, he devised plans for multi-stage rockets, living and working in weightlessness and even spacewalking.

As innovative as Tsiolkovsky's rocket designs were, they made use of a principle first stated in the seventeenth century by the English scientist Sir Isaac Newton – for every action there is an equal and opposite reaction. In a rocket, hot gases racing out of the engine nozzle (the "action") create a force that propels the rocket upward (the "reaction"). Although Tsiolkovsky never actually built a rocket himself, his ideas fuelled generations of space pioneers to come. His 1911 statement, now a mantra for advocates of space exploration, boldly stated humanity's imperative to leave its home planet: "The Earth is the cradle of the mind, but we cannot live forever in a cradle."

By the time of Tsiolkovsky's death, the technology for making his visions come true was already developing. In the United States a solitary inventor named Robert Goddard was developing the world's first liquid-fuelled rockets. Goddard knew that a rocket that burned liquid propellants would produce far more energy than one powered by solid fuel (for example, the black gunpowder used in fireworks). In addition, liquid-fuelled rockets can be turned off and re-started; they can also be throttled (that is, the flow of fuel can be controlled) – great advantages for a space vehicle. While Goddard's rockets were small, they were actually the direct ancestors of the mighty boosters that would one day send astronauts into space. However, no single person could solve the problems of large-scale rockets, and, ultimately, Goddard's main contribution to the history of rocket science was as an inspiration to those who followed him.

It took warfare to give rocketry the boost it needed. Just as aviation advanced during the First World War, so did rocket technology during the Second. In Nazi

Russian schoolteacher Konstantin Tsiolkovsky envisioned flights to the Moon in the early 1900s, half a century before his countrymen launched the first satellite.

Germany, a programme to develop a missile capable of striking targets in England and other Allied nations began under the leadership of a young engineer named Wernher von Braun. Inspired by German space visionary Hermann Oberth, von Braun had become obsessed with the idea of sending humans to the Moon and beyond. With the support of the Third Reich, von Braun and his team built what was at the time the largest and most powerful rocket in history, the V-2 "Vengeance" missile.

Standing 46 feet (14 metres) tall, the V-2 was powered by an engine that burned a mixture of alcohol and water with liquid oxygen to produce an awesome 56,000 pounds (250,000 newtons) of thrust. To make it work, von Braun and his team had to solve a host of technical problems, from creating a guidance system that would steer the missiles to their targets, to working with super-cold liquid oxygen, which exists at a temperature of -298° Fahrenheit (-183° Celsius). After numerous failures, they scored their first successful launch of the V-2 in October 1942. Travelling at five times the speed of sound, V-2s reached heights of up to 60 miles (96.5 kilometres), the first man-made objects to enter space. Ultimately, however, the V-2's greatest legacy was not as a weapon – though they were responsible for much death and destruction, many of the missiles failed to reach their targets and they failed also to frighten Great Britain out of the war – but as a forerunner of the rockets that launched some of the first satellites and astronauts into space.

Still, as the Second World War ended, the most important role for rockets was not in exploring space but as a means of engaging in a new kind of conflict. In the growing postwar tensions between the Soviet Union and the United States, world war gave way to cold war. Both sides studied the feasibility of developing the Intercontinental Ballistic Missile (ICBM), which could inflict nuclear destruction on an enemy a vast distance away in a matter of minutes by following a long ballistic arc that stretched into space and then descended toward its target. By any standard, the ICBM would be the ultimate weapon, so terrible that even its designers hoped it would never have to be used.

It would take a rocket many times more powerful than the V-2 to hurl a nuclear weapon weighing several tons over such distances. This meant developing more powerful engines and better guidance systems than those used in the V-2. Engineers also had to find a way to protect the warhead from the intense heat of its high-speed re-entry into the Earth's atmosphere. The difficulties of complex rocket engines and super-cold propellants were also present, but on a larger scale than any previous rocket effort. Most daunting of all was the need to counteract the irresistible pull of Earth's gravity to send the missiles from their launch sites to their distant targets. This meant that the ICBM had to be as lightweight as possible, without sacrificing power.

Few, however, doubted the importance of solving these problems. The Cold War was heating up, thanks both to the outbreak of the conflict in Korea in 1950 and the development of the hydrogen bomb by the two superpowers a few years later. In both countries, the pace of work intensified. By 1954 U.S. President Dwight Eisenhower had made the American ICBM, named Atlas, the nation's top defence priority.

Meanwhile, in the Soviet Union, a brilliant engineer named Sergei Korolëv was leading his own nation's ICBM project. Like Wernher von Braun, Korolëv dreamt of launching expeditions to the Moon and Mars. He had nurtured these dreams even as he toiled in Stalin's labour camps, where he was a political prisoner between 1938 and 1944. It was his work on wartime rocketry projects that won him his freedom. Had it not been for the Cold War, Korolëv's dreams of exploring space might have languished for decades; the Kremlin had no interest in his wild schemes. What mattered was the development of the ICBM – and Korolëv succeeded. His team produced the R-7 rocket, known to its creators by the Russian nickname *Semyorka* ("Little Seven").

Semyorka was designed to fire a hydrogen warhead at targets in North America, but Korolëv realized that an artificial satellite could hitch a ride on this military machine – and so could the space age. Both the Soviet Union and the United States had each planned to launch a satellite during the International Geophysical Year that began in mid-1957, however, it was Korolëv's R-7, tested for the first time in August, that gave the Soviet Union a chance to make history.

On the night of October 4, 1957, an R-7 stood ready for launch at a secret location on the steppes of Kazakhstan some 1,300 miles (2,000 kilometres) southeast of Moscow. At the nose of the R-7, concealed under a protective shroud, was a 22.8-inch (58-centimetre) sphere of polished aluminium containing a radio transmitter. If all went well, *Semyorka* would place this 184-pound (84-kilogram) creation into orbit around the Earth as the planet's first artificial satellite. While his launch team made their final preparations, Korolëv waited in a command bunker, trying to conceal his anxiety. He later told a colleague that he had been waiting for this day for most of his life.

Half an hour past midnight, brilliant flames erupted from the engines of the R-7. Moments later the great rocket rose into the night, lighting the steppes like an artificial sunrise. Apart from a couple of minor glitches, *Semyorka* functioned as designed, propelling Korolëv's satellite, called *Sputnik* (from the Russian meaning "fellow traveller"), beyond the atmosphere and into orbit around the Earth at a speed of 17,500 miles (28,000 kilometres) per hour. When the "beep-beep" of Sputnik's transmitter was received at tracking stations, Korolëv knew he had been successful. The space age had begun.

Robert Goddard (**above, left**), inventor of the liquid-fuelled rocket, in his workshop in Roswell, New Mexico in 1940.
In Germany, space visionary Hermann Oberth (**above right, left in this picture**) had many followers, including a young
Wernher von Braun.

The first modern rocket, the V-2, was developed as a weapon by Nazi Germany during the Second World War. Fired from mobile launchers in Germany and its occupied territories, V-2 missiles did severe damage to London (**above**) and other targets, but failed to influence the outcome of the conflict. After the war, captured V-2s found their way to scientists and engineers in the Soviet Union, and in the U.S., where many were launched from a test site at White Sands, New Mexico (**right**).

The success of the early Soviet space effort rested largely with Sergei Korolëv (**left**, with a future space traveller) who ran the main design bureau for spacecraft and boosters until his death in 1966. The first satellite, *Sputnik 1* (**below**) was launched on top of Korolëv's R-7 rocket on October 4, 1957 (**opposite**).

CHAPTER ONE
LEAVING THE CRADLE

The launch of *Sputnik* changed the world overnight – much to the surprise of Soviet Premier Nikita Khrushchev. At first, the Soviet leader was relatively unimpressed by the small, beeping sphere his rocket team had launched into space. That changed when he saw the world's reaction to the feat. Around the globe, reports of the first artificial satellite were greeted with amazement – or horror. In the United States, the news hit like a technological Pearl Harbor. Suddenly Americans could no longer view the Soviet Union as a backward nation incapable of competing on the world stage. Now it seemed the Soviets had mastered advanced rocket technology, something engineers in United States were still struggling with.

There was barely time to react to the first *Sputnik* when the Soviets launched a second in November 1957. *Sputnik 2* was not only more massive than its predecessor, it included the world's first space traveller: a dog called Laika. Sealed in her tiny pressurized cabin, Laika survived her trip into space, but perished within a few days due to a malfunction of the onboard environmental control system – a fact the Soviets did not reveal in their news reports. But the real significance of *Sputnik 2* was its size; it weighed more than 1,100 pounds (498 kilogrammes). If the Russian booster could launch such a huge satellite into orbit, surely it could do the same with a nuclear warhead that could be brought down on any American city. Suddenly the night sky seemed menacing.

Even aside from such dark possibilities, the American public felt the sting of having been beaten by its Cold War adversary. Before *Sputnik*, launching a satellite had not been very high on the list of the United States' national priorities; after the Soviet achievement it became an urgent matter. Success proved painfully elusive. In December 1957 the first American attempt to launch a satellite, called *Vanguard 1*, ended in disaster when its launcher fell back to earth moments after lift-off and exploded in a giant ball of flame. To make matters worse, the whole world knew about it; unlike the Soviet launches, which took place in secrecy, the American failure was broadcast live on television.

No one was more chagrined by this turn of events than Wernher von Braun. After the Second World War defeat of Germany, von Braun and many of his rocket team had turned themselves over to the U.S. Army, who brought them to America and put them to work developing missile technology, establishing a centre for this activity in Huntsville, Alabama. Von Braun had wanted to use his team's Jupiter C rocket to launch the American satellite, but had been shut out for political reasons; instead the navy's Vanguard rocket was chosen. His warnings to officials in Washington about the impending Soviet satellite launch had gone unheeded. But in the wake of *Sputnik*, von Braun suddenly gained credibility, and now, with Vanguard's failure, he and his team had their chance. On January 31, 1958, the Jupiter C launched a pencil-shaped container of scientific instruments called *Explorer 1* into orbit. America's first satellite distinguished itself by discovering a belt of radiation around the Earth, later named for one of *Explorer 1*'s developers, physicist James Van Allen. The mission's success meant that space exploration was no longer a one-nation effort; now it was a race between the world's two superpowers.

For now, the Soviets held the lead. In January 1959 they sent their *Luna 1* satellite racing past the Moon, the first artificial object to escape the Earth's gravitational pull. In September *Luna 2* did better by actually striking the lunar surface and in October, *Luna 3* circumnavigated the Moon and sent back the first images of the lunar farside, which is never visible from Earth. The Americans, meanwhile, suffered a series of embarrassing failures with their *Pioneer* spacecraft, which were designed to explore the lunar environment. Most failed to escape Earth orbit, and more than one blew up before reaching space. *Pioneer 4*, launched in March 1959, was able to fly past the Moon, missing it by 37,500 miles (60,000 kilometres).

By now, however, it was clear that the space race was about far bigger things. Everyone knew it was just a matter of time before human beings made their own space journeys, but how? In the United States, at the experimental flight-test centre at California's Edwards Air Force Base, a sleek, black rocket-powered airplane called the X-15 was being developed to fly to the edge of the atmosphere, 300,000 feet (90,000 metres) up. It would never be able to take a human into orbit, but designers were dreaming up a craft that would: *Dyna Soar*, a winged glider that could be launched from the top of a missile and land like an programme to put a man in space that came to be called Project Mercury. Unlike the X-15, the Mercury capsules were wingless and shaped something like a Styrofoam coffee cup. They were designed to be launched by a guided missile and then re-enter the atmosphere with a protective heat shield, splashing down in the ocean. By the autumn of 1959, the newly created National Aeronautics and Space Administration (NASA) had selected seven pilots as the nation's Mercury astronauts. Each hoped to make history's first piloted space mission, and they were not alone. In the Soviet Union, twenty pilots were training as cosmonauts to fly a spacecraft called *Vostok* that was being created by Sergei Korolëv's design bureau.

So, at the outset of the 1960s two nations were struggling to achieve something that seemed unthinkable to most people only a few years earlier. Apart from the technical challenges of sending a pilot safely into space and bringing them home again, there were the medical hazards. Beyond the atmosphere, a human being would need to be protected from radiation, extremes of heat and cold, and of course the vacuum of space. The intense forces of acceleration during launch and re-entry into the atmosphere were another concern. Some doctors warned that the body might not function normally in weightlessness. Others feared psychological disorientation that could hamper a pilot's ability to control a spacecraft. And then there was the simple fact that rockets had a disturbing habit of exploding.

With all of these concerns hanging over the first piloted Mercury flight, it's no wonder that American space officials decided to move cautiously. Mercury's first passengers were not human, but simian: there were two flights with rhesus monkeys, and two more with chimpanzees. Some rode the Redstone booster that would be used for Mercury's early, suborbital flights; others were launched aboard the Atlas ICBM that would send the pilots into orbit. By the spring of 1961 the astronaut selected for the first piloted Redstone mission, Alan Shepard, was more than ready to make the trip, but officials decided on one more test flight. For the United States, it would prove a fateful decision.

On the morning April 12, 1961, the first piloted spacecraft, *Vostok 1*, stood ready for launch on top of another of Korolëv's *Semyorka* rockets. Inside the *Vostok*'s spherical descent module, which was concealed inside an aerodynamic launch shroud, was a 27-year-old Russian pilot named Yuri Gagarin. If all went well, Gagarin would make a single orbit of the Earth before landing back in the Soviet Union.

One of the first space travellers, a dog named Belka (**opposite**), seen during her voyage in Earth orbit on board *Sputnik 5* in August, 1960.

Because no one could be sure how Gagarin would respond to weightlessness, *Vostok 1* was programmed to operate automatically, or by commands from mission control. Only in an emergency would Gagarin himself be allowed to take control – and then only after he had punched a three-digit code-number into a special locking device on the instrument panel. At 7:50 a.m. Moscow Time, Gagarin felt the rocket come to life and surge upward, and said over the radio, *"Payekhali!"* – "Lets Go!"

For the next 108 minutes, Gagarin made history. Inside *Vostok 1*'s cabin he made notes in his logbook – until his pencil floated away – ate and drank from special containers and observed the world passing below him. Only one moment of concern marred an otherwise perfect flight, when a cable linking his descent capsule with a conical instrument module failed to separate for several long minutes before re-entry. Not only did Gagarin suffer no ill effects, he returned to Earth exhilarated by his adventure, and as they had done with *Sputnik*, the Soviets once again scored a major public relations coup.

For the United States, the Gagarin flight was another stinging blow. It was also an embarrassment to the new president, John F. Kennedy, who was already dealing with the repercussions of a failed attempt to oust Cuban leader Fidel Castro. Like millions of Americans, Kennedy watched television on the morning of May 5, as Alan Shepard lay inside the spacecraft he had christened *Freedom 7* perched on a Redstone rocket. After a number of delays, the Redstone ignited and soared into the blue, sending Shepard on a 15-minute suborbital "hop" 116 miles (186 kilometres) above the Earth. As a technical achievement it did not match the Soviet orbital flight, but it boosted the spirits of the American public – so much so that Kennedy was moved to challenge the United States to a far greater goal in space.

Less than three weeks after Shepard's flight, addressing a joint session of Congress, Kennedy declared, "I believe this nation should commit itself to achieving the goal, before this decade is out, of landing a man on the Moon and returning him safely to the Earth." Space planners had done some detailed thinking about a lunar mission as early as 1960, but now they would actually have to make it happen. It would mean reaching a moving target that was 239,000 miles (384,500 kilometres) away in a mission requiring unprecedented accuracy. It meant building rockets more powerful than any yet devised and spacecraft so complex that their parts numbered in the millions. Even to some NASA managers, the President's challenge sounded fantastic. NASA had just fifteen minutes of spaceflight under their belts, and Kennedy was giving them less than ten years to get to the Moon.

For the time being, NASA's goals remained much simpler: to put a man in orbit. But that was something they could not rush into; in the summer of 1961 Gus Grissom repeated Shepard's suborbital mission. At the same time, as if to mock the Americans, the Soviets sent Gherman Titov into orbit for a full day. Finally, in February 1962, John Glenn rode an Atlas missile into space and circled the Earth three times. Glenn's flight achieved Project Mercury's central objective, but it also gave NASA its first crisis.

Sometime before reentry, ground controllers scanning telemetry from *Friendship 7* saw an indication that its heat shield had detached. If so, the fiery passage through the atmosphere would be fatal to Glenn. After deliberating, they came up with a strategy centred on *Friendship 7*'s package of retrorockets. The package, which was attached to the craft's heatshield by straps, was designed to be jettisoned before re-entry. But mission control told Glenn to leave it attached, thinking it would help hold the heatshield in place. The controllers would have to wait to find out if their plan was successful; as *Friendship 7* slammed into the upper atmosphere a sheath of glowing, ionized gas surrounded the spacecraft, blocking communications with Earth.

As it turned out, the indication of a loose heatshield was erroneous, and Glenn's re-entry was uneventful – except for the flaming chunks of the retropackage that Glenn saw flying past his window. When he regained radio contact, Glenn told mission control, "My condition is good, but that was a real fireball, boy." Glenn's safe return was a blaze of glory for the American public, whose adulation approached that given to Charles Lindbergh following his solo transatlantic flight in 1927.

Three more *Mercury* astronauts followed in Glenn's orbital footsteps, culminating in May 1963 with Gordon Cooper's 33-hour mission. But the Soviets continued to outdistance American progress in space. In June 1963 they launched *Vostok 5* with Valiery Bykovsky on a mission lasting nearly five days. Two days into his flight, Bykovsky was joined in orbit by the first woman in space, Valentina Tereshkova, in *Vostok 6*. By the time the duo returned to Earth, Soviet cosmonauts had logged a total of almost 16 days in space, as compared with just over two days for American astronauts. However, both sides had shown that humans could survive trips into space. Both knew that enormous challenges lay ahead, but they also knew, as one NASA controller wrote when John Glenn reached orbit, "We are through the gates."

The first attempt by the United States to launch a satellite (**opposite**) ends in a very public defeat as the rocket carrying *Vanguard 1* explodes seconds after lift-off on December 6, 1957. Weeks later, on January 31, 1958 success finally arrived with the launch of *Explorer 1* (**above**) perched on a Jupiter C missile. Celebrating the feat by holding aloft a model of their creation are, (**above right from the left**), Jet Propulsion Laboratory director William Pickering, University of Iowa scientist James van Allen and designer of the Jupiter-C rocket, Wernher von Braun.

The far side of the Moon, which is never seen from Earth, was first glimpsed by *Luna 3*, a Soviet probe that flew around the Moon on October 7, 1959. In this image the lunar far side occupies the right-hand portion. A prominent crater with a dark floor and bright central peak, visible on the lower right, was named for Russian space-pioneer Konstantin Tsiolkovksy.

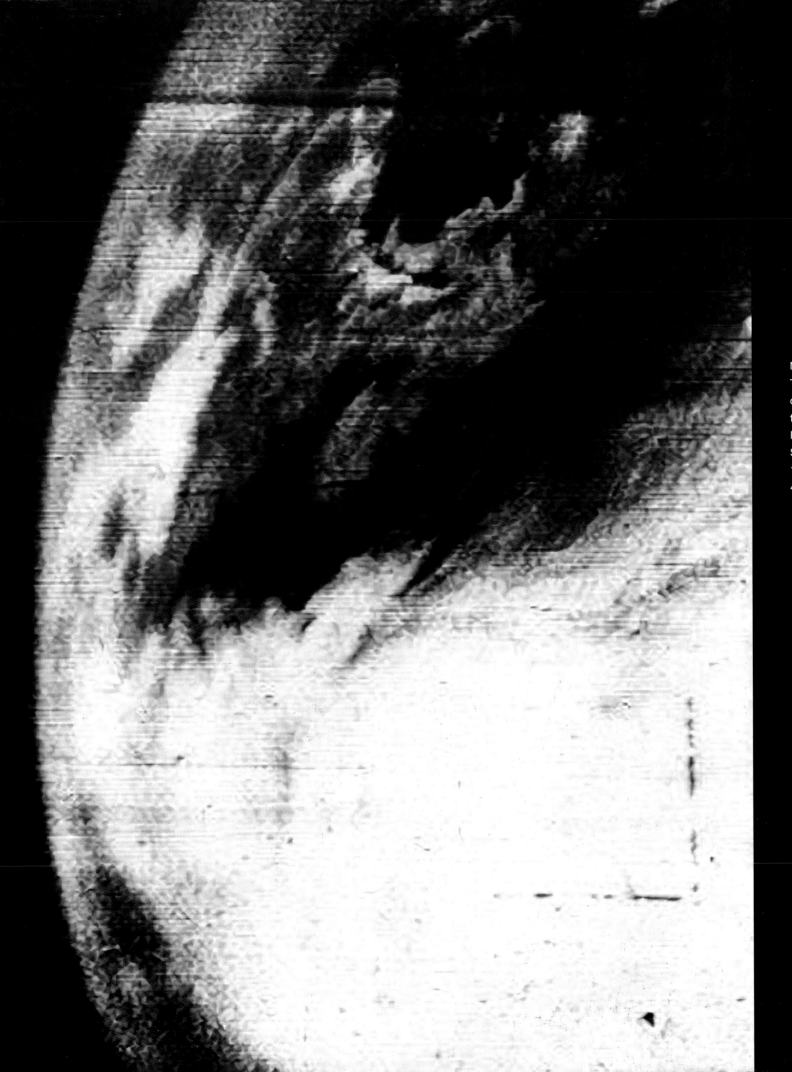

New eyes on our world.
The first television image
of Earth from space was
transmitted by Television
Infrared Observational
Satellite (Tiros 1) on April 1,
1960 from a height of
450 miles (724 kilometres).

NASA's original seven astronauts pose in their space suits. Clockwise from top left: Alan Shepard, Virgil "Gus" Grissom, Gordon Cooper, Walter "Wally" Schirra, Donald "Deke" Slayton, John Glenn and Scott Carpenter.

An image showing six of the early Soviet cosmonauts (opposite), including the first man in space, the first woman in space and the first spacewalker. From the left: Pavel Popovich, Alexei Leonov, Yuri Gagarin, Pavel Belyayev, Valentina Tereshkova and Andrian Nikolayev.

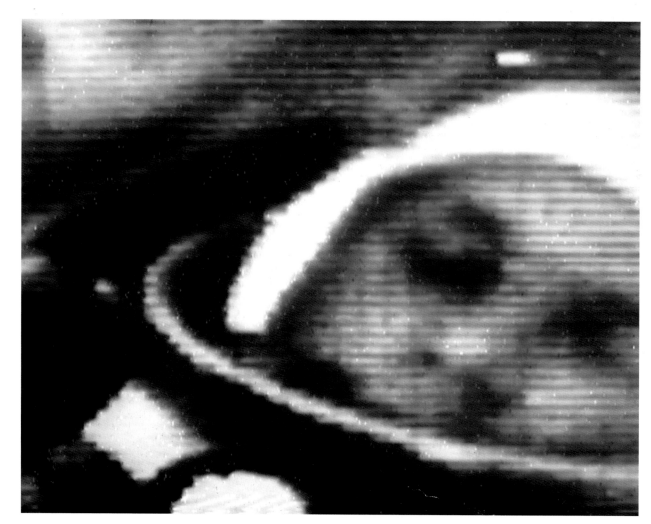

Yuri Gagarin was a 27-year-old fighter pilot with relatively little experience when he became the first man to fly in space on April 12, 1961. Clad in his space suit, Gagarin is seen during training (**left**), with Korolëv, the architect of the Soviet space programme (**opposite, left**) and in the cabin of his *Vostok 1* spacecraft (**opposite, right**) in a view televised to Earth during his one-orbit flight. During the mission Gagarin wrote notes in a logbook (**above**) until his pencil floated away.

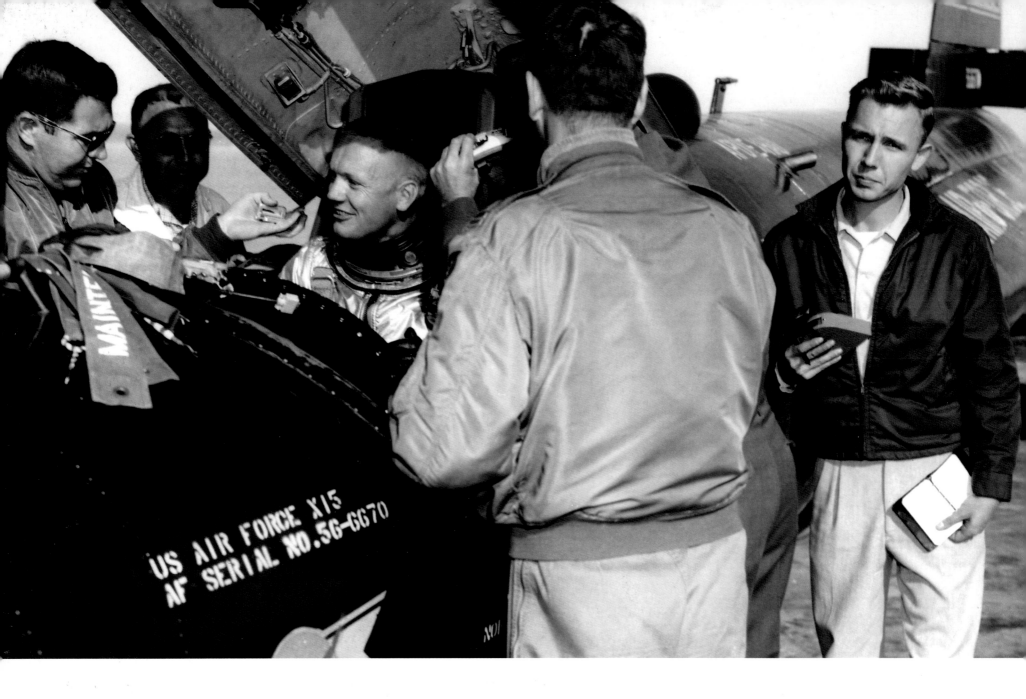

The rocket-powered X-15, a joint creation of NASA and the U.S. Air Force, Navy and North American Aviation, flew at more than six times the speed of sound and reached heights in excess of 350 thousand feet (110 thousand metres), at the edge of space. One of its pilots was a young civilian named Neil Armstrong (**opposite**), seen here after a flight in 1960, nine years before he commanded the first lunar landing mission.

Alan Shepard lies in the tiny cabin of his Mercury spacecraft *Freedom 7* on the morning of May 5, 1961, as technicians prepare to install the craft's side hatch. After hours of delays, Shepard lifted off on top of a Redstone booster (**opposite**) to become the first American in space. His 15-minute suborbital hop took him to a height of 116 miles (187 kilometres) before splashdown near Bermuda.

Following the trail blazed by Yuri Gagarin, John Glenn rode an Atlas booster into Earth orbit on February 20, 1962 (**opposite**). Inside the cabin of his Mercury craft, *Friendship 7* during his four-and-a-half hours in space (**above**), Glenn took photographs through a small overhead window.

Gordon Cooper strides across the deck of the carrier USS *Kearsarge* after his 34-hour Mercury mission on May 16, 1963.

Cooper's feat barely measured up to the missions of Soviet cosmonauts, who by this time were logging multi-day missions. Valentina Tereshkova (**above**, in training) became the first woman in space with an orbital voyage lasting nearly three days aboard *Vostok 6* in June, 1963.

CHAPTER TWO
THE RACE FOR THE MOON

As 1964 opened America was a nation in mourning. The young president who symbolized a new energy for a new decade, John F. Kennedy, had been brutally assassinated in November 1963 as his motorcade rode through Dallas. However, Kennedy's boldest initiative, the challenge to put a man on the Moon during the 1960s, was very much alive. At NASA the basic plan for the lunar mission had been worked out. It called for two spacecraft: a command ship that would ferry three astronauts to and from the Moon and a bug-like lunar lander that would take two of them down to the surface and back to lunar orbit. At the space agency, and at aerospace contractors around the country, engineers were busy creating these craft and preparing for their voyages.

Before any astronaut could fly them, however, NASA would have to answer some crucial questions. Could astronauts survive a lunar round trip lasting a week or even longer? Could they work in the vacuum of space, protected only by a pressurized space suit? Could they master the intricate ballet called orbital rendezvous, in which one spacecraft crew chases down another – a key step in the lunar flight plan? These unknowns were the domain of Project Gemini, a series of two-man missions designed to serve as NASA's training ground for the Moon.

Although it bore a superficial resemblance to Mercury, the Gemini spacecraft was a major advance over its predecessor. What Gemini lacked in creature comforts – if anything its crew cabin was slightly more cramped than Mercury's – it more than made up for in capabilities. Designed with the pilot in mind, Gemini would be capable of changing the shape and orientation of its orbit around the Earth. It would feature the first computer onboard a spacecraft and would even be able to control its re-entry through the atmosphere to reach a pre-chosen splashdown point. But even as Mercury veteran Gus Grissom and rookie John Young prepared for the first piloted Gemini mission, the Soviets once again took centre stage. In October 1964 cosmonauts Vladimir Komarov, Boris Yegorov and spacecraft design engineer Konstantin Feoktistov spent a day in orbit inside a new craft called *Voskhod 1*. Some at NASA wondered whether Voskhod was the Soviet version of the Apollo command ship; if so, the Soviet Union had leapfrogged them in the Moon race before the first piloted Gemini could even get airborne.

In reality, Voskhod had nothing to do with going to the Moon. The brainchild of Sergei Korolëv, Voskhod was created from Vostok hardware with one main goal, to upstage Gemini. The Voskhod crew cabin was a Vostok descent module in which the single ejection seat had been replaced by three couches; Komarov, Yegerov and Feoktistov flew without the benefit of space suits in order to save room in the cramped cabin. Feoktistov, a member of Korolëv's design bureau, had objected to the plan – until Korolëv promised him a seat on the flight.

In March 1965 *Voskhod 2* reached orbit, this time with two space-suited cosmonauts. Early in the flight Alexei Leonov entered a collapsible airlock, sealed himself off from mission commander Pavel Belyayev, and emerged into the void. For about ten minutes, Leonov floated outside *Voskhod 2*, secured to the ship by an umbilical line. He almost didn't make it back inside; his pressurized suit ballooned so badly in the vacuum of space that he only managed to squeeze back through the airlock by letting some of the oxygen out of his suit, risking serious injury from decompression sickness. After Leonov finally rejoined his commander, exhausted but unharmed, the Soviets reported on his feat – minus its near-disastrous conclusion – and scored yet another public relations victory in the space race.

However, Voskhod proved to be a flash in the pan. Korolëv had more Voskhod missions in preparation, but by this time he was under pressure to get on with the Russian Moon programme which, unknown to the West, was getting a rather late start. There would be no more Soviet human spaceflights in 1965, and none the following year; and as the piloted Gemini flights got underway – Grissom and Young reached orbit just a few days after *Voskhod 2*'s return – NASA began to make great strides with a spectacular series of missions.

Every two months, a new Gemini team rocketed into orbit with an ambitious flight plan. In June 1965 *Gemini 4*'s Ed White improved on Alexei Leonov's endeavour by walking in space for some 20 minutes, propelling himself for part of that time using a nitrogen-powered manoeuvring gun. Two months later *Gemini 5*'s Gordon Cooper and Pete Conrad racked up eight days in orbit, setting a new space endurance record. In December, Frank Borman and Jim Lovell almost doubled that figure, staying in orbit for 14 days aloft *Gemini 7,* battling systems malfunctions, cramped living conditions and boredom. The highlight of their mission came when they were joined in orbit by *Gemini 6*'s Wally Schirra and Tom

Stafford, who steered their spacecraft to within a foot (0.3 metres) of their companions' in history's first space rendezvous. By the end of 1965 the Soviets knew their lead had evaporated; the Americans were now the leaders in human spaceflight experience.

Farther out in space, the United States was also taking the lead in robotic exploration of the heavens. The *Mariner 2* probe made the first reconnaissance of another planet in December 1962 when it flew by cloud-hidden Venus and revealed that its surface temperature was hot enough to melt lead. In July 1965 *Mariner 4* raced past Mars and sent back the first close-up images of the red planet, showing a desolate, cratered surface, but the world receiving the most intense scrutiny was the Moon, whose mysterious surface was of obvious interest to Apollo planners. NASA's lunar assault began with a series of "crash-landers" called Ranger which sent back close-up images before impacting the surface. In 1966 the first of the Surveyor soft-landers arrived (a few months after the Soviet *Luna 9* probe successfully touched down and transmitted panoramic views of dust and rocks). Between 1966 and 1968 five Surveyors made successful landings and sent back thousands of detailed images; some of them provided data on the Moon's chemical composition and even probed the surface directly with a mechanical scoop. And five Lunar Orbiter probes, equipped with telescopic cameras, mapped the Moon from above in stunning detail.

As the Moon became less mysterious, so did the challenges of going there. The Gemini missions continued into 1966, as astronauts accomplished the first space dockings, undertook longer spacewalks, refined the art of orbital rendezvous and made the first pinpoint landings of the U.S. space programme. To be sure, there were close calls, none more dire than *Gemini 8*'s wild tumble through space, caused by a stuck manoeuvring thruster that almost killed Neil Armstrong and Dave Scott. There were harrowing moments too for Gemini's spacewalkers, who struggled to master the strange difficulties of working in the three-dimensional ice rink called weightlessness. *Gemini 12*'s Buzz Aldrin finally demonstrated that the spacewalk gremlins had been tamed in a record five hours of excursions in November 1966. By the time he and his commander Jim Lovell splashed down, ending the Gemini programme, NASA was ready to reach for the Moon.

However, as the United States and the Soviet Union were both about to learn, winning that prize would come at a painful cost. On January 27, 1967, the crew of the first piloted Apollo mission, a planned two-week "shakedown cruise" in Earth orbit, died when a flash fire swept through their sealed command module during a ground test. In the wake of the tragedy it became clear that the fire had resulted from a series of design flaws, including the use of pure oxygen at high pressure inside the crew cabin on the launch pad. Amid criticism from the press and the public, NASA focused on recovering from the fire and making the end-of-the-decade lunar deadline.

The Soviets suffered their own tragic loss in April 1967 when Vladimir Komarov piloted the new Soyuz spacecraft. Roughly equivalent in capabilities to Gemini, Soyuz was designed to carry up to three people into Earth orbit, and variants of the craft were being created for the lunar programme. Unlike Gemini, however, Soyuz was meant to be flown mostly by automatic pilot or ground control, an approach that added to its complexity. There had been malfunctions during several unpiloted test flights, and the decision to launch Komarov was made over the resistance of some space planners. During the mission, Komarov had been scheduled to link up with a second, unpiloted Soyuz, but that plan was cancelled after one of *Soyuz 1*'s solar panels malfunctioned. Then came problems with the craft's automatic control system. Komarov was ordered back to Earth and managed to steer *Soyuz 1* through a difficult manual re-entry. Unfortunately, the craft's parachutes did not deploy properly and *Soyuz 1* slammed into the ground at high speed, killing Komarov instantly. Some Soviet planners had wanted to send a pair of cosmonauts around the Moon in time for the fiftieth anniversary of the Russian Revolution in October 1967; that hope died with Komarov.

The Americans' hopes, meanwhile, got a boost – quite literally – in November with the first unpiloted test flight of the giant *Saturn V* Moon rocket. Together with the Apollo spacecraft this behemoth, whose creation was guided by Wernher von Braun, stood 363 feet (111 metres) above its launch pad; to this day the *Saturn V* remains the largest and most powerful rocket ever built. At lift-off the Kennedy Space Center shook with the thunder of the rocket's first-stage engines, which delivered an awesome 7.5 million pounds (33.5 million newtons) of thrust. During the 11-hour mission, designated *Apollo 4*, the Saturn showed itself to be a worthy flying machine – welcome news in the worst year of NASA's brief history.

By the spring of 1968, however, there were new worries. The Apollo lunar module was facing one technical obstacle after another, and planners realized it would not be ready for its piloted debut, slated for the end of the year. At the space centre in Houston, the brilliant engineer and Apollo manager George Low came up with a stunning plan to get back on track. If no serious problems cropped up on *Apollo 7*, the first piloted mission in October, then *Apollo 8* could fly around the Moon in December, without a lunar module. The risks of the proposed mission – the astronauts would have no "lifeboat" if their command ship failed – were outweighed by the benefits; in one step NASA could test the basic procedures for getting to the Moon and getting back. October brought good news, as *Apollo 7* sailed through an 11-day Earth-orbit test flight with no significant problems, validating the new, fireproof command module. Only a head cold suffered by mission commander Wally Schirra marred the otherwise successful mission. Schirra and his crew even found time to entertain Earthbound viewers with a series of in-flight television broadcasts, at one point holding up a sign that read, "Keep Those Cards and Letters Coming In" in a nod to old-time radio shows. The most exciting result of *Apollo 7* was NASA's go ahead for *Apollo 8*'s lunar voyage, as Frank Borman, Jim Lovell, and Bill Anders prepared to become the first humans to leave Earth orbit.

In the Soviet Union, Alexei Leonov also longed for that honour. Together with flight engineer Oleg Makarov, he was training to fly a Soyuz variant called *Zond* in a figure-of-8 loop around the Moon sometime in the fall of 1968. While not as complex as *Apollo 8*'s planned lunar orbit flight, the Soviet circumlunar mission would nevertheless represent a significant achievement – and another public relations victory in the Moon race. But there were problems with *Zond* on two unpiloted test flights, and Soviet space planners, only too mindful of the death of Vladimir Komarov, denied permission for Leonov and Makarov to launch. They could only watch as *Apollo 8* readied for its own lunar journey.

Early in the morning of December 21, Borman's crew rode their *Saturn V* into Earth orbit; less than two hours later they re-ignited the third-stage engine to head out of Earth orbit on a trajectory for the Moon. For 66 hours they crossed the vast trans-lunar gulf, watching their home world shrink until the men could cover it with an outstretched thumb. Then, just past midnight on December 24, they fired *Apollo 8*'s main engine and slipped into orbit around the Moon. Through the command module's windows, Borman, Lovell and Anders saw an ancient and alien landscape, so pockmarked with meteorite craters that it could have been a battlefield. For 20 hours the men circled this desolate world, taking photographs, making navigation sightings, and beaming television pictures to Earth – including a stirring reading of the first 10 verses of Genesis, set against a backdrop of the primordial lunar landscape, in a Christmas Eve message that would be remembered by all who witnessed it. For the astronauts, the most electrifying sight of the voyage was the Earth rising beyond the Moon's bleached and lifeless horizon; indeed, all three felt that they had come all the way to another world to discover the one they had left behind.

Getting back to that bright blue Earth meant traversing 235,000 miles (380,000 kilometres) of space and flying a precise path through the atmosphere. Too steep an entry and the command module would be torn to pieces by the forces of deceleration. Too shallow and the craft would skip off the atmosphere like a stone skimming the surface of a pond, and Borman's crew would die a lonely death in the depths of space. The moment of truth came on December 27 in the pre-dawn darkness over the Pacific, as *Apollo 8* slammed into the atmosphere, precisely on schedule and on target. Inside the command module the men were treated to an eerie lightshow as friction with the atmosphere created a sheath of glowing gas around the spacecraft. Finally, three main parachutes deployed and *Apollo 8* splashed down safely, bringing history's first Moon voyagers home. In retrospect, the mission also took the wind out of the sails of the Soviet Moon programme; they knew now that only a miracle would land cosmonauts on the Moon before their American rivals.

NASA barely had time to celebrate the success of *Apollo 8*'s bold adventure when it was time to prepare for the next steps: *Apollo 9*'s full-up test of the Apollo spacecraft – including the lunar module – in Earth orbit, and *Apollo 10*'s "dress rehearsal" for the landing mission in lunar orbit. Each of these missions was so complex that it made *Apollo 8* look simple by comparison. Remarkably, each went off nearly flawlessly, opening the way for *Apollo 11* to meet John F. Kennedy's challenge in July.

Commanding the mission was Neil Armstrong, veteran of *Gemini 8*'s harrowing tumble in space. It would be up to Armstrong to steer the lunar module to a safe touchdown on the Moon's Sea of Tranquillity, while Gemini spacewalker Buzz Aldrin served as co-pilot. Fellow Gemini veteran Mike Collins would fly the command module to and from the Moon, and for some 30 hours alone in lunar orbit while his crewmates travelled to and from the surface. Of course, the highlight of the mission would be a Moonwalk by Armstrong and Aldrin lasting about 2½ hours. Aside from the historical significance of human footsteps on another world, the flight would give scientists their first samples of the Moon's rocks and dust to study – if the astronauts returned safely. In his own mind, Armstrong felt sure he and his crew would survive the mission but he believed the chances for a successful lunar landing were only fifty-fifty.

On July 19, 1969, after following the path traced by *Apollo's 8* and *10*, Armstrong, Aldrin and Collins arrived in lunar orbit. The next day, Armstrong and Aldrin transferred into their lunar module *Eagle*, checked its systems, and undocked, leaving Collins in the command module *Columbia*. After firing *Eagle*'s engine to lower their orbit, Armstrong and Aldrin coasted down to 50,000 feet (15,000 metres), and re-ignited the engine to begin their final descent. On the way down, the *Eagle*'s computer became overloaded and threatened to abort the landing; that was averted thanks to some swift analysis by experts in mission control. No sooner had this crisis passed, however, than Armstrong discovered that the onboard guidance system was steering *Eagle* toward a stadium-sized crater ringed with car-sized boulders. With not far to go, Armstrong took semi-manual control and began hunting for a safe landing spot.

In the final moments of the descent, a storm of Moon dust kicked up by *Eagle*'s descent engine blurred Armstrong's view of the surface and confused his perception of motion, making his piloting task even tougher than he had anticipated. Knowing that the fuel supply was dwindling, Armstrong guided *Eagle* down the last few feet until, suddenly, a blue light on the instrument panel signalled that three lunar contact probes had touched the surface. As *Eagle* settled gently onto the Moon, Armstrong shut down the engine, and then all was still. After a congratulatory handshake from Aldrin, Armstrong announced to the world that they had achieved the dream of ages: "Houston, Tranquillity Base here. The *Eagle* has landed."

After a check of *Eagle*'s systems, Armstrong and Aldrin got the go-ahead from mission control to stay on the Moon and after eating a meal (beef and potatoes, butterscotch pudding, brownies and grape punch) the pair began preparations to go outside. Like modern-day knights in armour, they sealed themselves in pressurized space suits that contained layers of protection against extremes in temperature, and even micrometeorites from space. Underneath, each man wore special long-underwear that kept its wearer cool by circulating water through a network of tiny tubes. There were lunar overshoes designed to give steady footing on lunar dust, and a reflective visor to screen out harmful sunlight and lunar gloves with an outer layer of woven metal fibre. Finally, a backpack supplied breathing oxygen, cooling water and radio communications, via *Eagle*, with Earth.

On Sunday evening, Houston time, the men vented the oxygen from the cabin of the lunar module, and opened the small, square front hatch. Squeezing out of the cabin on his belly, Armstrong emerged into the vacuum. After pulling a cord to deploy a black-and-white television camera, he carefully descended nine rungs of *Eagle*'s ladder and reached the foil-covered footpad. For a few moments he surveyed his surroundings, and then Armstrong carefully placed his left boot onto the ancient lunar ground, paused, and spoke for the history books: "That's one small step for [a] man, one giant leap for mankind."

Minutes later, Aldrin followed his commander to the surface, and for almost two hours the pair explored a dusty area of the lunar surface. They raised the American flag, deployed a pair of experiments, and collected rock samples while an estimated 600 million people – a fifth of the world's population, watched on live television or listened on radio. History's first walk on another world went without a hitch, and the pair climbed back into their lander, their space suits somewhat dirty with lunar dust, and settled in for the night. Meanwhile, in lunar orbit, Mike Collins attended to *Columbia*'s systems and tried, unsuccessfully, to spot *Eagle* on the surface using his onboard sextant. Collins' real concern,

of course, was that his crewmates make it safely back to him; that would depend on *Eagle*'s ascent rocket. If it failed, Armstrong and Aldrin would be stranded on the Moon with no hope of rescue.

Midday on July 21, Houston time, Armstrong and Aldrin stood ready for their lunar lift-off. Neither man had slept well in the lander's cabin, but that hardly mattered: both were intensely focused on what came next. Right on schedule, the ascent engine came silently to life, propelling the men into the black lunar sky like a super-fast elevator ride. Minutes later they were back in lunar orbit, steering a path toward their companion in *Columbia*. From then on, *Apollo 11* went precisely according to plan, ending three days later with splashdown in the Pacific. The American space programme had met Kennedy's challenge with a few months to spare.

But even as NASA looked ahead to the next lunar landing, *Apollo 12*, there was uncertainty about what the future would bring for the U.S. space programme. Planners envisioned a fantastic array of projects including a reusable space shuttle, a permanently manned space station, bases in lunar orbit and on the surface and even human voyages to Mars. It would be up to a new president, Richard Nixon, to decide the fate of those visions, even as Americans prepared to return to the Moon.

"I felt absolutely free, soaring like a bird … as though I had wings

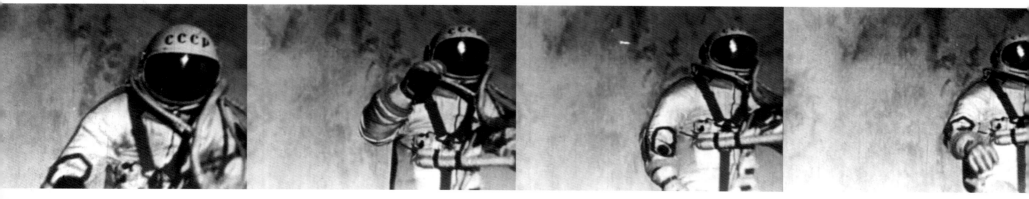

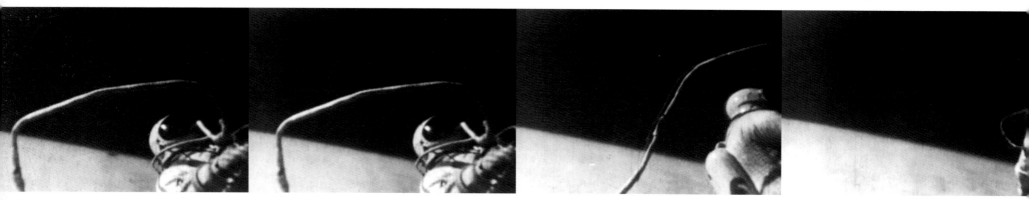

as though I was flying by my own efforts." ALEXEI LEONOV

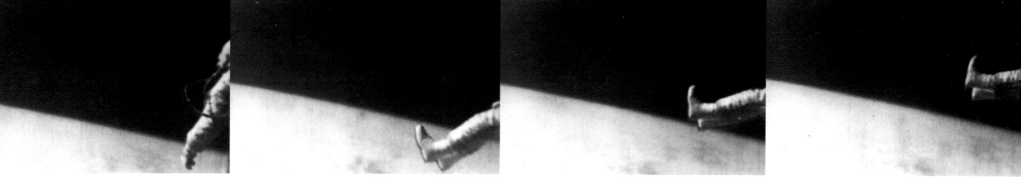

Alexei Leonov floats outside the *Voskhod 2* spacecraft in history's first spacewalk, March 18, 1965. These scenes were recorded by a movie camera attached to the craft's exterior.

Gemini 3 leaves the launchpad on March 23, 1965, carrying Gus Grissom and John Young on Gemini's piloted debut.

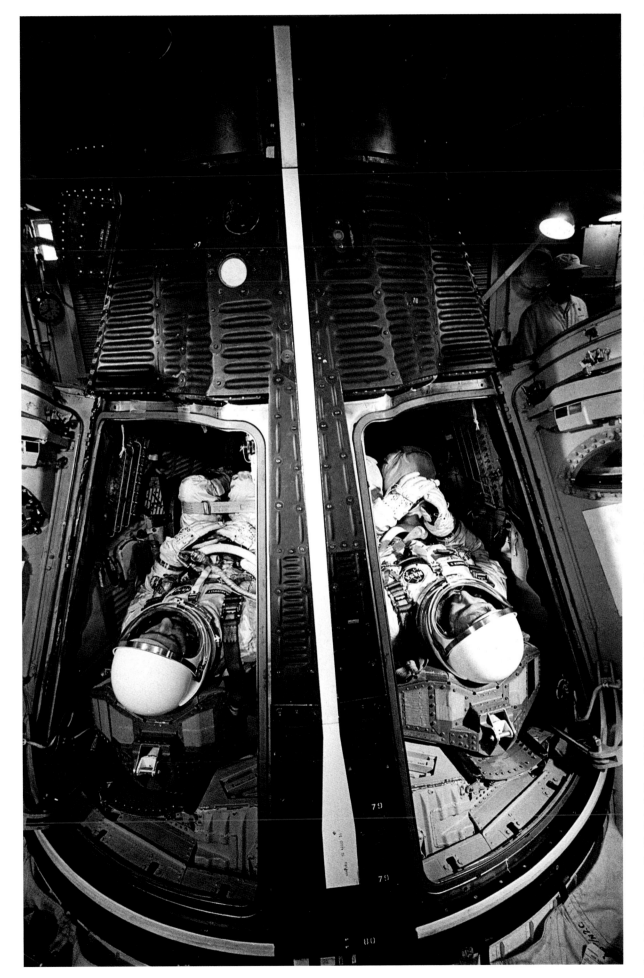

Gemini 4 astronauts Jim McDivitt (**left**) and Ed White await the closing of the craft's hatches before launch on June 3, 1965. Each man lies in an ejection seat designed to be used in case of an emergency during the ascent to orbit – a situation every astronaut hoped to avoid.

A hundred miles (160 kilometres) above the Earth, Ed White floats at the end of a 25-foot (7.6 metre) umbilical cord during the first U.S. spacewalk on June 3, 1965. In his right hand, White holds a nitrogen-powered manoeuvring gun, with a Nikon camera attached.
On his chest is an emergency oxygen pack; his gold-plated visor is designed to protect against the unfiltered sunlight of space. White's excursion in the void lasted 23 minutes.

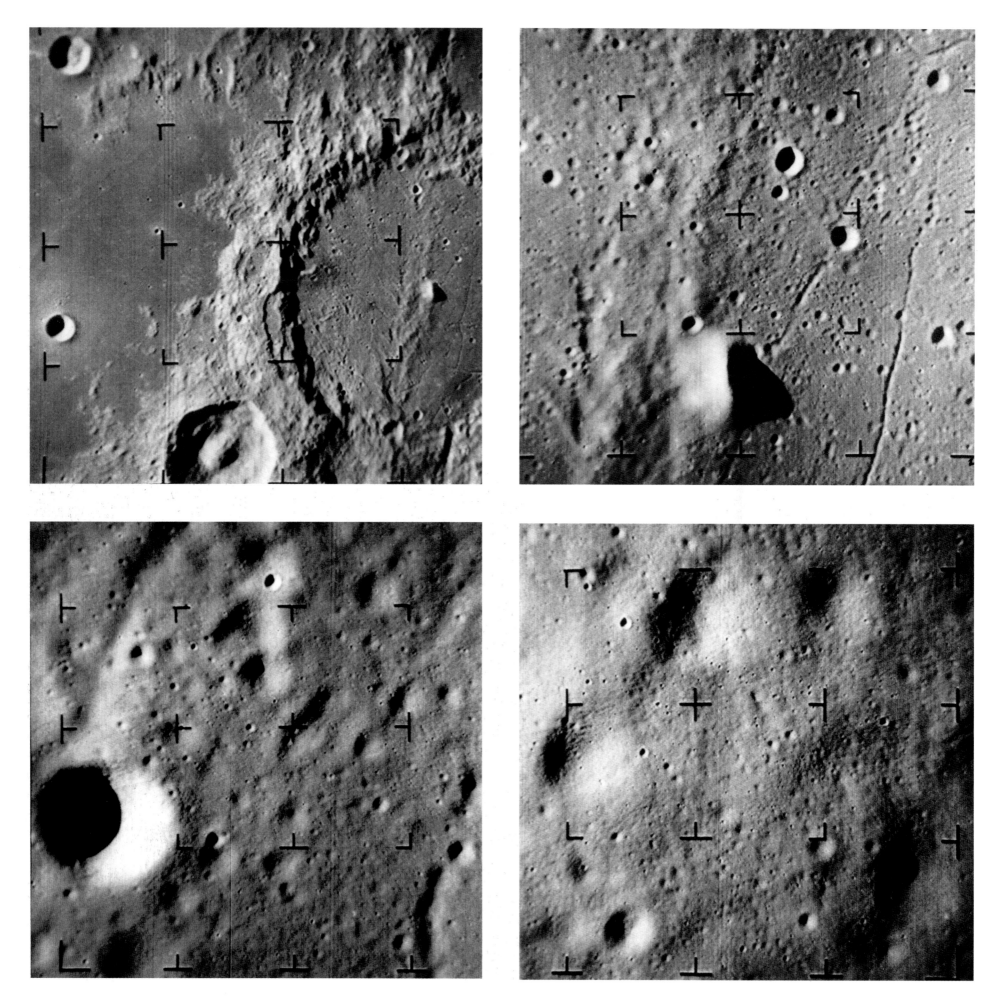

Chapter Two

Preceding pages. Humans got their first close-up look at the moon from the Ranger series of space probes, which sent back television pictures before crashing into the lunar surface. This sequence of images records *Ranger 9*'s plunge into the 74-mile (119-kilometre)-diameter crater Alphonsus on March 24, 1965. **On page 60, clockwise from upper left:** 18 minutes before impact, altitude 1,550 miles (2,494 kilometres), image is 141 miles (227 kilometres) across; 39 seconds before impact, altitude 58 miles (93.3 kilometres), image 37 miles across (59.5 kilometres); 8 seconds before impact, altitude 16 miles (26 kilometres), image is 7 miles (11.3 kilometres) across; less than 3 seconds before impact, altitude 4.5 miles (7.25 kilometres), image 2 miles (3.2 kilometres) across. **On page 61:** at an altitude 9.5 miles (15.3 kilometres), some 5.5 seconds before impact, *Ranger 9*'s telephoto camera recorded this view, which shows craters as small as 50 feet (15.25 metres) across.

This page. The first close-up image of Mars, transmitted by *Mariner 4* during its flyby of the Red Planet on July 14, 1965. **Opposite:** *Mariner 4*'s best image of Mars, showing a 96-mile (154.5 kilometre) diameter crater that was later named to honour the spacecraft.

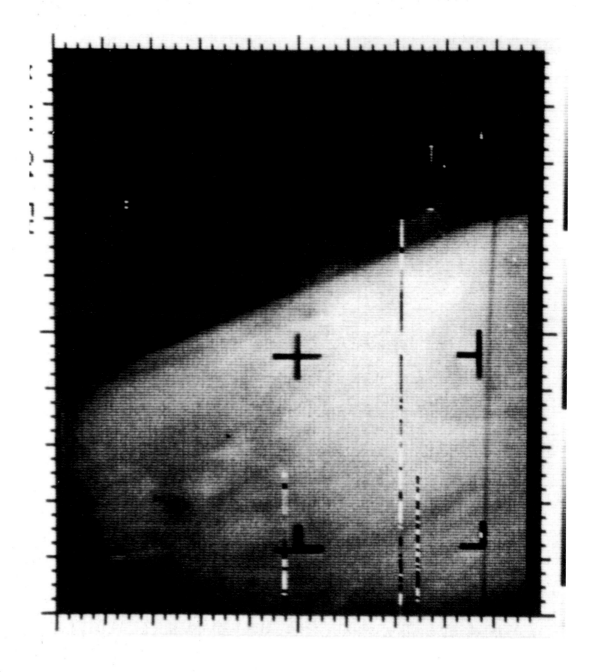

By 1966 the Soviets and Americans were dispatching soft-landing craft to the Moon, which transmitted the first images from its surface. The Soviet *Luna 9* alighted on the Moon's Ocean of Storms on February 3, 1966 (**above**); the U.S. *Surveyor 1* touched down near the crater Flamsteed on June 2, 1966 (**opposite**) and sent back thousands of images, including this one of its own shadow. One of *Surveyor 5*'s three footpads rests on the powdery soil of the Sea of Tranquillity (**below**) in September 1967.

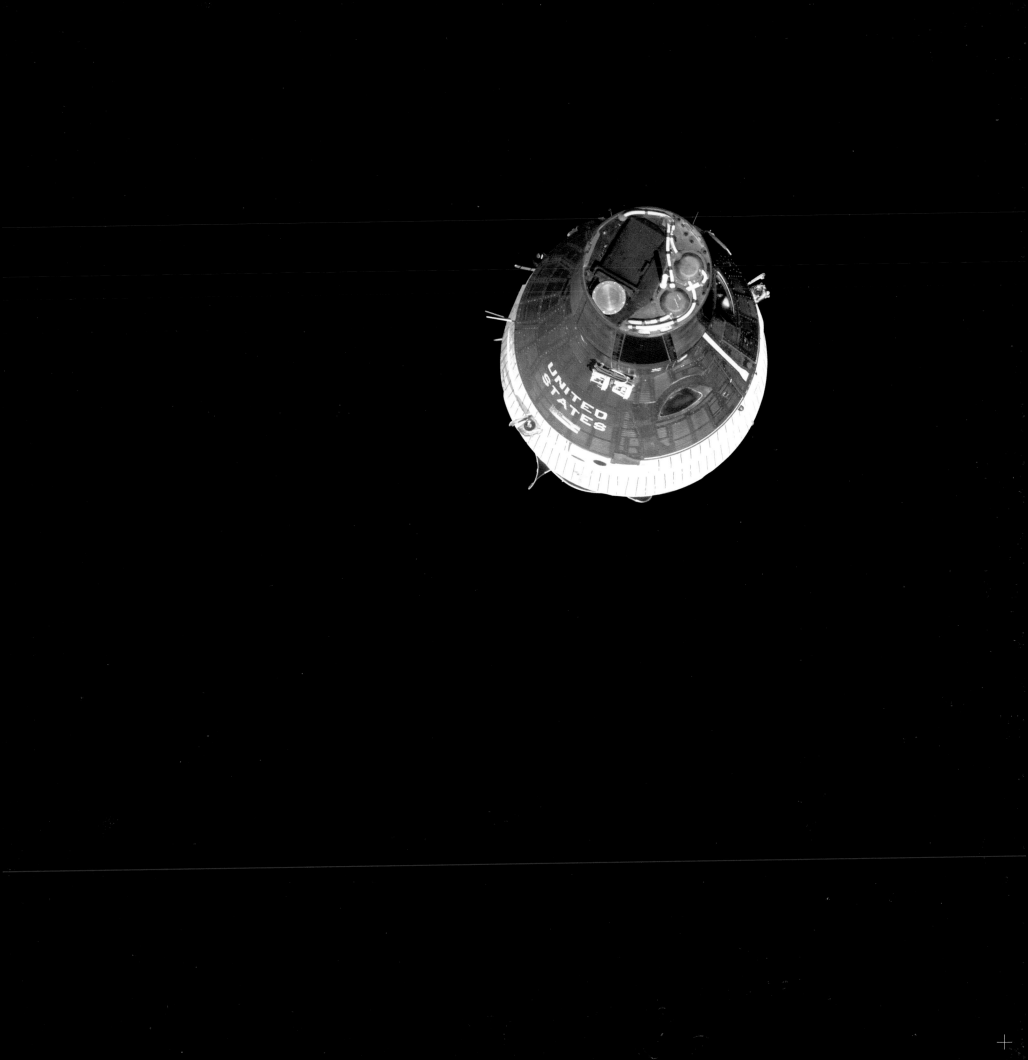

Preceding pages. History's first space rendezvous – a crucial milestone for the planned lunar missions of Project Apollo – was accomplished by Wally Schirra and Tom Stafford in *Gemini 6*, who chased down their orbiting colleagues Frank Borman and Jim Lovell in *Gemini 7* on December 15, 1965. Stafford took these views of *Gemini 7* during the orbital *pas de deux*.

The strain of a two-week space marathon shows on the face of *Gemini 7* commander Frank Borman (**above**), performing a medical experiment designed to test his vision in weightlessness. Glowing exhaust streams from the engine of an Agena target rocket (**opposite**) docked to the *Gemini 10* spacecraft in July 1966. Aboard *Gemini 10*, John Young and Mike Collins fired the Agena's engine to raise the height of their orbit to a record altitude of 473 miles (761 kilometres).

Sitting astride the nose of *Gemini 11* (**above**), Dick Gordon struggles through a difficult spacewalk, recorded by an onboard movie camera on September 13, 1966. Gordon became exhausted while trying to fix an experimental tether from the craft to the attached Agena target rocket. The following day Gordon and mission commander Pete Conrad fired their Agena's engine to climb to 850 miles (1,368 kilometres) above the Earth. On the way up, Gordon photographed India and Sri Lanka (**below**).

Standing in the open hatch of *Gemini 12* on November 12, 1966, Buzz Aldrin snaps a self-portrait during one of three spacewalks he made during the mission. Aldrin logged a total of 5½ hours outside, proving that astronauts could perform useful work in the vacuum of space without becoming exhausted.

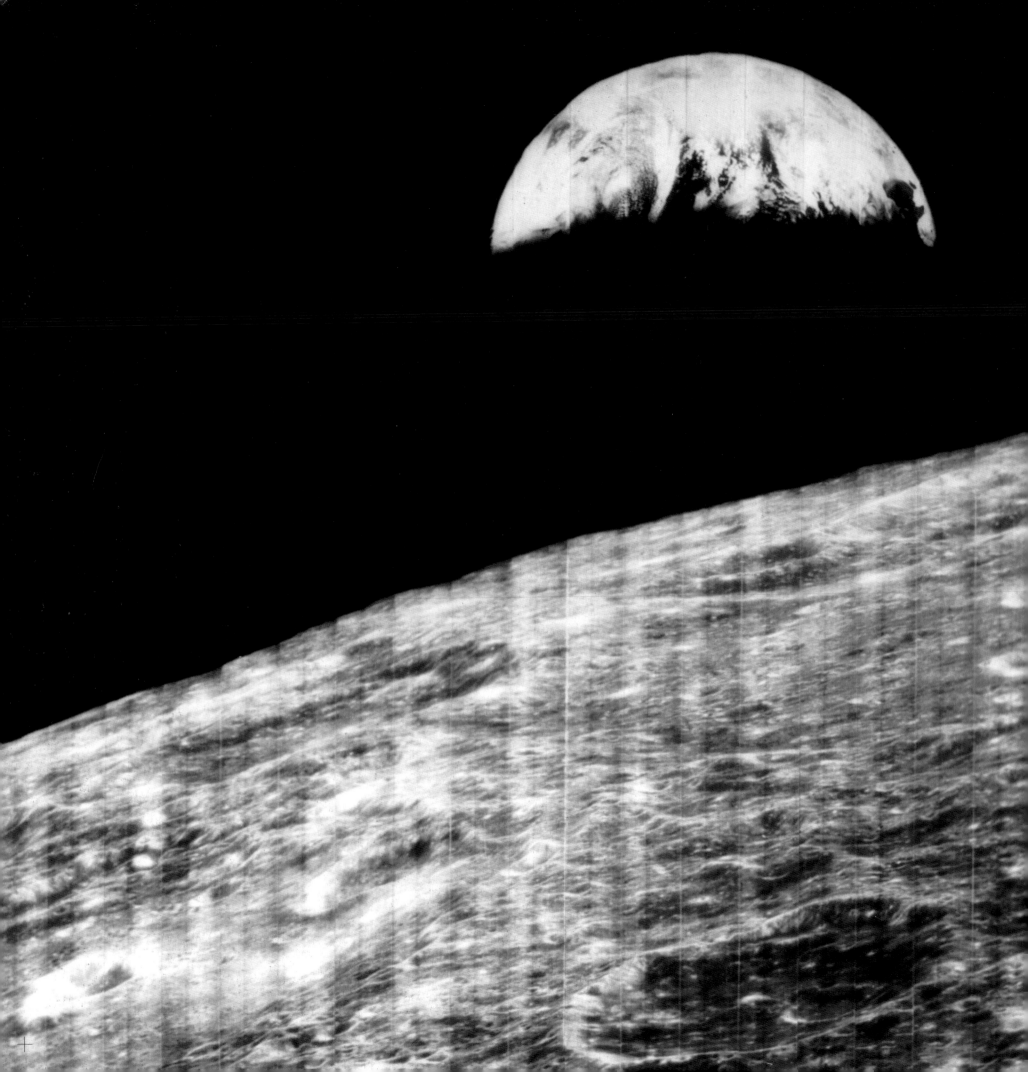

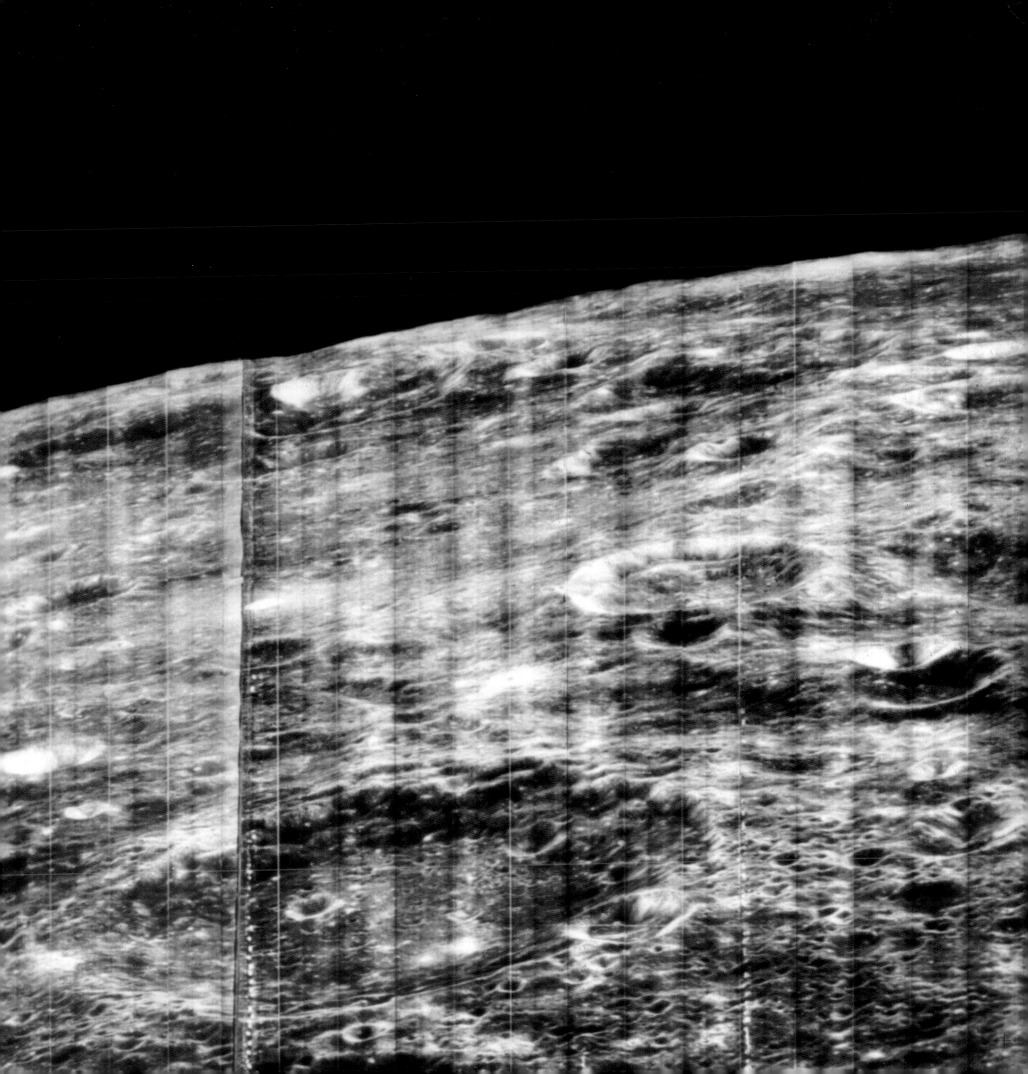

Tragedy struck the U.S. space programme on January 27, 1967 when the crew of the first piloted Apollo mission (**right, from the left**), Gus Grissom, Ed White, and Roger Chaffee, were killed during a practice countdown when a flash fire erupted inside their spacecraft. A day after the fire, the cabin of *Apollo 1* (**above**) was a mass of charred wreckage.

Preceding pages. Orbiting the Moon on August 23, 1966, the unpiloted *Lunar Orbiter 1* took the first photograph of the Earth rising above the lunar horizon. The scene was recorded on photographic film developed on board the spacecraft, which were scanned and the resulting images transmitted to Earth.

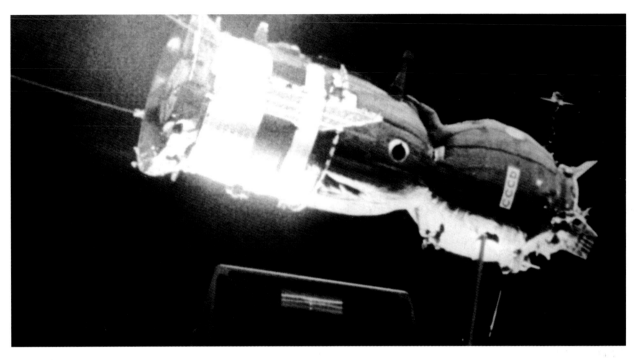

The first man to die during a spaceflight, cosmonaut Vladimir Komarov (**left**), perished on April 24, 1967 when the parachutes on his *Soyuz 1* spacecraft failed to open properly. Soyuz craft (**above**, seen in 1975) were designed for a range of missions in Earth orbit and variants of the craft were being prepared to carry men around the Moon.

Preceding pages. The first *Saturn V* rocket rises from Pad 39A at the Kennedy Space Center on the morning of November 9, 1967 on an Earth-orbit test flight. Flight-testing the giant rocket "all-up" rather than in pieces – a gamble by NASA managers – probably saved months in the race to land humans on the Moon.

Vindicating the re-designed, fireproof Apollo command module, *Apollo 7* was a complete success on the craft's maiden, Earth-orbit mission in October 1968. During the 11-day voyage Walter Cunningham makes notes (**opposite**) and does a somersault in zero-g (**above**).

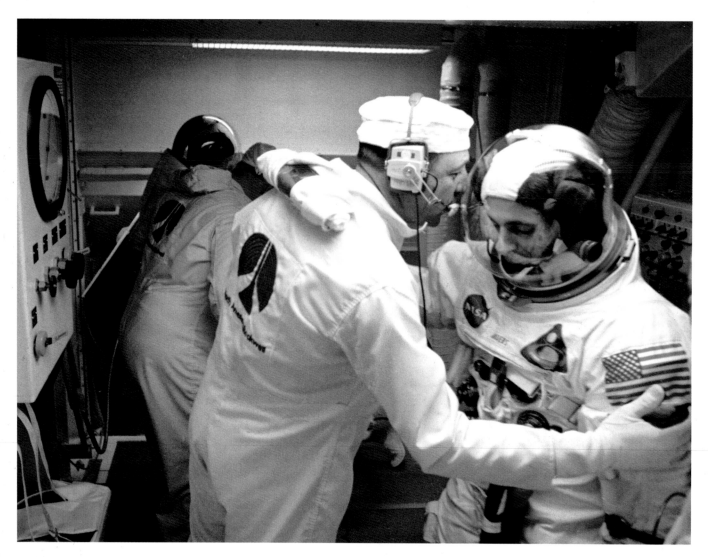

On the morning of December 21, 1968, Bill Anders bids farewell to a technician before boarding the *Apollo 8* command module for the first flight around the Moon. Hours later (**opposite**), on a Moonward course, Anders and his crewmates, Frank Borman and Jim Lovell, became the first humans to see the Earth as a planet. In this view North and South America lie along the boundary between day and night.

Looking down on the far side of the Moon, the *Apollo 8* astronauts saw a barren, crater-pocked landscape; the largest crater in this view is 20 miles (32 kilometres) across. As they drifted over the near side once more, the astronauts witnessed the electrifying sight of their home planet rising above the bleached lunar horizon, as photographed by Anders (**opposite**).

Overleaf. The ambitious *Apollo 9* mission in Earth-orbit included a spacewalk on March 6, 1969 by Russell "Rusty" Schweickart to test the lunar spacesuit and backpack; during his excursion Schweickart photographed crewmate Dave Scott (**left-hand page**) standing in the open hatch of the command module *Gumdrop*. On March 7 Schweickart and mission commander Jim McDivitt undocked from Scott and flew the lunar module *Spider* on the lander's first manned test (**right-hand page**).

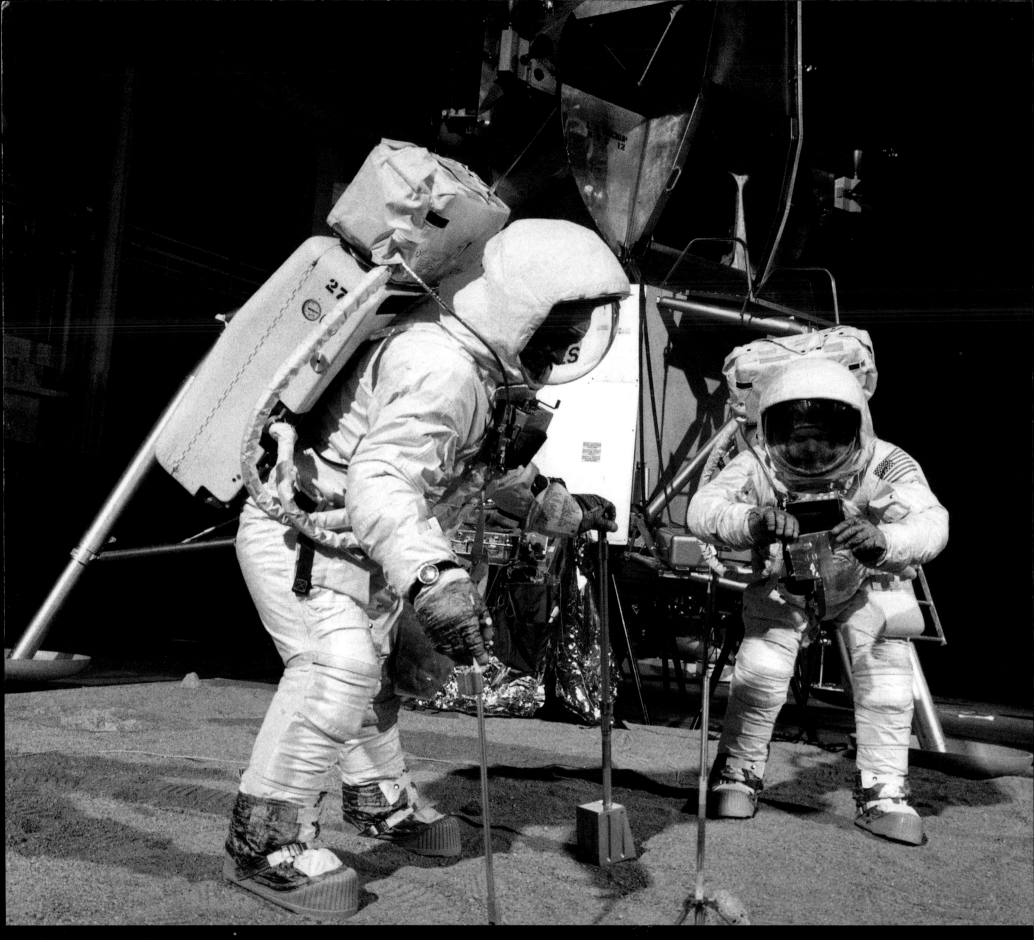

Preceding pages. The *Apollo 10* command ship *Charlie Brown* and lunar module *Snoopy* after undocking in lunar orbit on May 22, 1969. In a dress-rehearsal for the lunar landing, Tom Stafford and Gene Cernan flew *Snoopy* to within 9 miles (14.5 kilometres) of the Moon's surface before rejoining John Young in *Charlie Brown*.

April 22, 1969. Inside a cavernous training building at the Manned Spacecraft Center in Houston, *Apollo 11* astronauts Neil Armstrong (**above, right**) and Buzz Aldrin practice tasks for history's first Moonwalk.

On July 11, just five days before launch, Armstrong (**left**) and Aldrin practice for their descent to the Moon inside the lunar module simulator at the Kennedy Space Center in Florida.

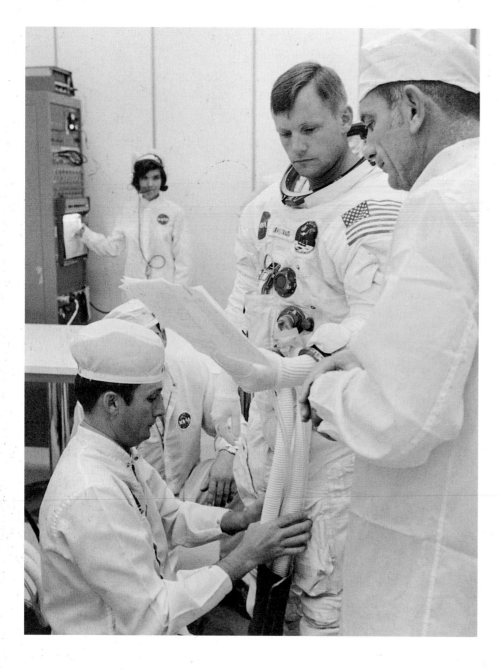

Neil Armstrong is seen having a pre-launch conversation with his boss, astronaut Donald "Deke" Slayton, while suiting up on the morning of July 16, 1969 (**above left**). He then leads his crew to their spacecraft (**above right**). Trailing an immense column of flame *Apollo 11* speeds out of the atmosphere (**opposite**).

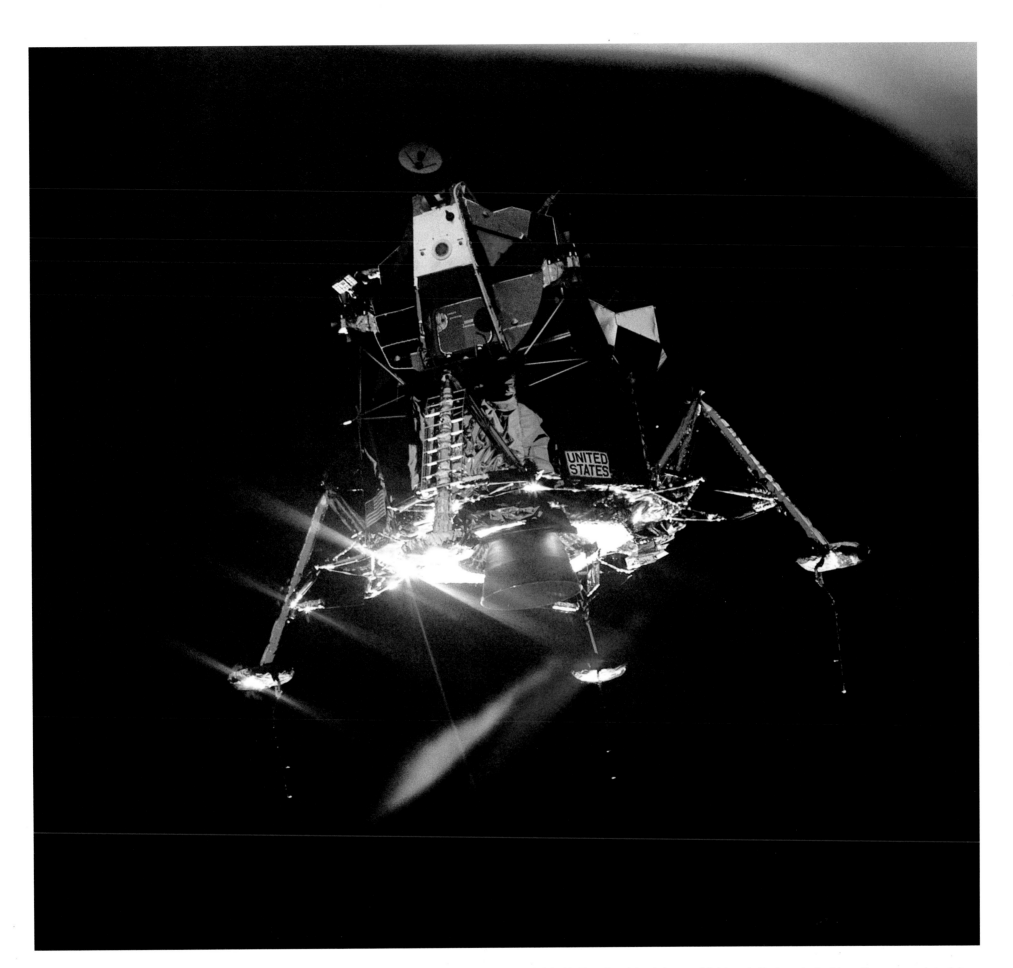

Inside the lunar module *Eagle* (**opposite**) Buzz Aldrin inspects the lander during the Moonward voyage on July 18. Two days later, in lunar orbit (**above**), *Eagle* undocked from the command module *Columbia*, to leave Mike Collins on a 22-hour solo voyage.

Standing in *Eagle*'s front footpad, Armstrong takes his first step onto the Moon at 9:56 p.m. Houston time, July 20, 1969 (**above**). The moment was transmitted live by an onboard television camera to an Earthbound audience estimated at 600 million people.

Scenes from the Moonwalk, which was recorded by a movie camera placed in *Eagle*'s right-hand window. From the top, Armstrong's first step; Armstrong using an equipment conveyor to bring down a Hasselblad camera; Armstrong collecting an initial sample of lunar dust and rock fragments; Armstrong (**left**) and Buzz Aldrin raising an American flag; Aldrin (**left**) and Armstrong walking toward the lander after taking a phone call from U.S. President Richard Nixon.

Aldrin poses for Armstrong's camera (**opposite**); his visor reflects one of *Eagle*'s landing legs, the Moonscape and the photographer.

Among the signs of human presence left by the first Moonwalkers were a plaque on *Eagle*'s front landing leg bearing the inscription "Here men from the planet Earth first set foot upon the Moon, July, 1969 A.D. We came in peace for all mankind." An hour into the Moonwalk, Aldrin photographed his own bootprint in lunar dust (**opposite**). With only the rain of micrometeorites from space to erode them, the astronauts' footprints should remain on the lunar surface for millions of years.

CHAPTER THREE
CHANGING HORIZONS

In 1970 the Moon looked no different, seen from Earth, than it had for millions of years but few who gazed up at it could deny that it was forever changed: humans had walked there. The first lunar landing was the greatest engineering feat of the millennium and even though *Apollo 11* had won the Moon race, NASA's lunar explorations were just getting into high gear. Four months after Armstrong and Aldrin's visit to the Sea of Tranquillity, *Apollo 12*'s Pete Conrad and Alan Bean had explored the Ocean of Storms. In the first pinpoint lunar landing, they'd touched down within walking distance of the unpiloted *Surveyor 3* probe, showing that astronauts could now visit places of interest to geologists, places where the secrets of the Moon's origin and evolution might be gleaned. For scientists, the Earth's satellite was now a world ripe for exploration.

But in the spring of 1970, NASA got another reminder that sending men to the Moon was a dangerous adventure. On April 11 there was an explosion aboard the Moon-bound *Apollo 13* that crippled the command module and plunged the crew, Jim Lovell, Jack Swigert and Fred Haise, into a battle for survival some 200,000 miles (320,000 kilometres) from home. Quick thinking by the astronauts and teams of experts in mission control allowed Lovell's crew to use their attached lunar lander as a lifeboat to get back onto a course for Earth. When Lovell, Swigert and Haise splashed down safely on April 17, their ordeal ended – and so did the rescue effort that many consider NASA's finest hour. Now the agency focused on figuring out what had gone wrong, and resuming lunar exploration. Apollo would continue and, at NASA, there were still hopes that the Moon would be a gateway to even greater projects in space.

Not long after *Apollo 13*, however, it became clear that these dreams would not come to pass. To the American public, lunar missions no longer held the same fascination, and other concerns had begun to crowd Apollo from the national arena, including civil rights, the environment, and above all the war in

Vietnam. NASA realized that neither the Nixon administration nor Congress wanted to pay for the great visions offered by space planners; there would be no bases on the Moon, no piloted missions to Mars – at least, not anytime soon. Instead, NASA faced a trend of budgetary belt-tightening; by the summer of 1970 it had cancelled the three final lunar landings to save money and hardware for other programmes.

If Apollo's end was premature, it was also spectacular. The final three missions were scientific expeditions, outfitted with extra supplies that allowed a pair of astronauts to spend three days on the Moon. With improved spacesuits that offered greater mobility, they took Moonwalks lasting up to seven hours, a full working day of exploration. Best of all, these teams brought along their own transportation: a battery-powered Lunar Rover that allowed them to travel across the landscape, even onto the sides of mountains.

It was on one of these peaks that *Apollo 15*'s Dave Scott and Jim Irwin discovered a sample of the Moon's primordial crust in July 1971. The sample, nicknamed the Genesis Rock, proved to be 4.5 billion years old, dating back almost to the birth of the Moon itself. Such moments of discovery were shared live on television with scientists and spectators alike, thanks to the Rover's colour television camera, which could be controlled remotely from Earth.

Geologists longed to see one of their own explore the Moon, and they got their wish on the final lunar landing, *Apollo 17*, in December 1972. Geologist-astronaut Jack Schmitt and his commander, veteran Gene Cernan, explored the Taurus-Littrow valley, as deep as the Earth's Grand Canyon. While they roamed the surface, crewmate Ron Evans explored the Moon from orbit, just as his predecessors had, aided by a battery of high-powered cameras and other sensors. Finally, when Cernan, Evans and Schmitt splashed down in the Pacific on December 19, the Apollo programme was over. No one could know then that these would prove to be the last lunar explorers of the twentieth century.

Things might have gone differently if the Moon race had not ended so decisively. But as it happened, the Soviet Union's dreams of sending men to the Moon – already embattled when *Apollo 11* landed there – were all but dead by the time of *Apollo 17*. The only Soviet visitors to the Moon were robots, including a pair of rovers called *Lunakhod*, and a trio of automated sample return missions. The giant N-1 booster, cornerstone of the Soviet Moon landing programme, failed in four launch attempts; its development was finally cancelled in 1974, officially ending the country's manned lunar effort. Soviet space planners deeply felt the absence of the late Sergei Korolёv's drive and inspiration; surely he would have looked to Mars as a destination for cosmonauts.

Instead, the Soviets focused their efforts on another goal, much closer to home: a space station in

Earth orbit. In April 1971, in time for the tenth anniversary of the first spaceflight by Yuri Gagarin, they launched the *Salyut 1* space station. The name, which translates as "salute" was a nod to the late cosmonaut, who had perished in a plane crash in 1968. Consisting of a single, cylindrical module, *Salyut 1* contained space for three cosmonauts to live and work in Earth orbit for weeks at a time.

Salyut 1's first inhabitants, a trio of cosmonauts, spent a record-breaking 22 days aboard the station, but their achievement was overshadowed by the flight's tragic end. Inside their Soyuz taxi spacecraft, the cosmonauts perished when an air valve opened before re-entry, venting the cabin atmosphere into space. The craft re-entered automatically but when ground crews opened the hatch, they found the bodies of the crew inside. The tragedy forced a hiatus in the Soviet space station programme, which would continue in 1973.

The United States, meanwhile, launched its own space station in May 1973. Called *Skylab*, it was constructed from leftover Apollo hardware, including the third-stage booster of a *Saturn V* Moon rocket. With as much living space as a small house, *Skylab* was also designed to be an orbiting research facility, but the station's life began traumatically when a protective micrometeoroid shield was torn off by aerodynamic stresses during launch, taking one of the station's two power-producing solar panels with it. The second panel, snared by a piece of debris from the torn shield, had failed to extend. Low on power and overheating in the Sun's scorching rays, the unoccupied station was threatened even before its mission could begin. It was up to *Skylab*'s first crew, led by Apollo Moonwalker Pete Conrad, to rescue the programme.

Soon after arriving onboard the station in late May, Conrad and his crew extended a reflective sunshield above the exterior of the station, allowing temperatures to come down. Later, Conrad and physician-astronaut Joe Kerwin made a daring spacewalk to free the stuck solar panel, wielding a pair of long-handled cutting shears; they succeeded, restoring *Skylab* to full working order. The next *Skylab* crews extended the space endurance mark even further: Alan Bean led his crew through a 56-day visit, followed by the final crew, led by rookie Gerry Carr, for a 94-day stay that ended early in 1974. *Skylab*'s scientific bounty included thousands of images from a high-resolution solar telescope, new photos of the Earth and medical data on the human body's response to long-duration spaceflight.

But like Apollo, *Skylab*'s potential exceeded its lifetime. Plans for a second *Skylab*, one that could be refurbished, were scrapped as NASA concentrated its efforts on its next major goal in space, a reusable space shuttle. Approved in 1972, the Shuttle would be launched like a rocket and land like an airplane.

With it, NASA hoped to lower the cost of access to Earth orbit and open up the space frontier for countless science, engineering and even manufacturing activities. But the Shuttle would not be ready to fly until 1978 at the earliest. By the time *Skylab* had finished, NASA was planning only one more human spaceflight – one that would have been unthinkable just a few years before.

Slated for launch in the summer of 1975, the Apollo-Soyuz Test Project was to be the first docking between spacecraft of the United States and the Soviet Union. Born of the growing détente between the two Cold War rivals, it was designed to pave the way for future cooperation in space. The highlight of the mission came on July 17, 1975, as Apollo veteran Tom Stafford piloted the final Apollo command ship toward a rendezvous with a Soyuz taxi commanded by spacewalker Alexei Leonov. The two ships met with a firm thump and Leonov called out in English, "We have capture …. Okay, Soyuz and Apollo are shaking hands now."

Little more than two hours later, high over France, hatches between the two craft were opened and Stafford and Leonov had a true handshake in space that began two days of joint operations. History's first international space mission went so well that one NASA flight director publicly stated, "I wish there was another one of these flights. We've gone to all this trouble to learn how to work with those people … I could run another Apollo-Soyuz or another joint anything with a heck of a lot less fuss than it took to get this one going." However, Apollo-Soyuz was a unique, one-time venture.

Meanwhile, there were other adventures underway, much farther from home, as the United States and Soviet Union dispatched a fleet of robotic explorers to other worlds of the solar system. The inner planets Mercury and Venus were visited by *Mariner 10* in 1973 and 1974 and the hellish Venusian surface was reached by twin Soviet landers *Venera 9* and *10* in 1975; the probes survived long enough to send back the first images from the surface of another planet.

Perhaps the most intriguing world, Mars, was also the object of the greatest number of missions. Despite a daunting number of failures in both United States and Soviet Mars efforts – the Soviet Union launched no less than 16 Mars missions between 1960 and 1973, none of which were successful – the planet had lost none of its lure. The missions that did succeed only raised more questions, in particular, *Mariner 9*'s orbital survey that began in 1971. The Mars revealed by *Mariner 9*'s cameras was not the Moon-like expanse glimpsed by earlier probes, but a complex world brimming with geologic wonders. There were giant, apparently extinct volcanoes, the largest of which towered five times as high as Mount Everest. An enormous network of canyons, stretching the length of the continental United States, scarred the surface. And most intriguing, winding channels resembling terrestrial dry river valleys snaked across the Martian deserts.

Scientists realized that Mars had once been a very different world, one which was geologically alive.

But what about life itself? Could living things, even microscopic ones, have evolved on Mars? Late in the summer of 1975 two robotic explorers called *Viking 1* and *2* left Earth in hopes of finding out. Each Viking lander was outfitted with a miniature laboratory in which samples of Martian dust would be analyzed for their chemical composition and for any signs of microbial activity. Few scientists held out hope that the Martian surface – bathed in harmful ultraviolet radiation and apparently devoid of liquid water – would harbour life, but the possibility was too enticing to ignore.

Viking 1 touched down successfully on July 20, 1976, seven years to the day after Armstrong and Aldrin walked on the Moon. *Viking 2* followed in September. During years of operation on Mars, the landers sent back images of rock-strewn, rust-coloured deserts under a salmon-hued sky. Neither craft turned up convincing evidence that life exists on Mars – but that disappointment stood in contrast to the landers' scientific bounty, including surprising data on the presence of highly oxidizing compounds in the Martian dust. The twin Viking orbiters, circling Mars, mapped the planet in unprecedented detail.

Even Mars was not the limit for NASA. Thanks to a once-in-a-lifetime alignment of the outer planets, it was possible to send a single probe to as many as four of these giant worlds, in far less time than normally required. The key was a technique called gravity assist, in which a spacecraft uses the gravity of one planet like a slingshot to send it on to another world. *Mariner 10* had demonstrated this technique during its flyby of Venus, which redirected the craft to Mercury. In the outer solar system, it meant using the gravity of Jupiter to head to Saturn, then to Uranus and Neptune. Planners called it the Grand Tour.

There were two trailblazers for the Grand Tour, the *Pioneer 10* and *11* probes, which flew past distant Jupiter in 1973 and 1974, sending back images and revealing new details about the planet's powerful magnetic field. *Pioneer 11* went on to a rendezvous with Saturn in 1979 but the main events, so to speak, were *Voyager 1* and *2*, which left Earth in 1977 on their own planetary odysseys, outfitted with high-resolution cameras and a slew of other sensors. In 1979 each craft made a reconnaissance of Jupiter and its family of moons, then, using the powerful Jovian gravity, each sped onward to Saturn, as *Pioneer 11* had done before them. Their explorations, and those of other planetary explorers, would continue into the 1980s, as a backdrop to new eras in human spaceflight.

Apollo 12 lifting off in a rainstorm on November 14, 1969 with Pete Conrad and his crew. Flying through a thundercloud, the spacecraft was struck by lightning, temporarily knocking out the craft's electrical system. Six days later, in the second of *Apollo 12*'s two Moonwalks, Conrad inspects the unpiloted *Surveyor 3* probe (**right**), which had landed on the slopes of a crater on the Moon's Ocean of Storms in 1967. On the crater's rim, 600 feet (183 metres) away rests the *Apollo 12* lunar module *Intrepid*, which Conrad piloted with Alan Bean in the first pinpoint lunar landing.

Unaware of the crisis awaiting him and his crew in space, *Apollo 13* commander Jim Lovell peers out from his spacesuit a few hours before launch on April 11, 1970.

On April 13, 1970, when *Apollo 13* was 200,000 miles (320,000 kilometres) from home, an oxygen tank aboard the service module exploded, crippling the command module. At Mission Control (**above**), astronauts and flight controllers tracked the growing crisis. Using their attached lunar module as a lifeboat, Lovell and Jack Swigert (**right**, photographed by crewmate Fred Haise) endured a long and chilly journey back to Earth. Shortly before reentry the astronauts photographed their cast-off service module (**opposite**), where a gaping hole reveals the damage caused by the explosion.

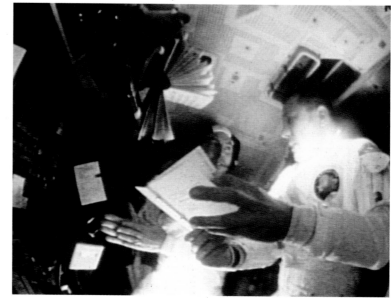

Lovell smiles from a so-called
Billy Pugh net as he is lifted to the
recovery helicopter on April 17, 1970;
the command module is visible
at the left edge of the photo.

Exhausted from their ordeal (**left**), the *Apollo 13* astronauts emerge onto the deck of the carrier USS *Iwo Jima* while in mission control (**below**), flight director Gene Kranz smokes a celebratory cigar.

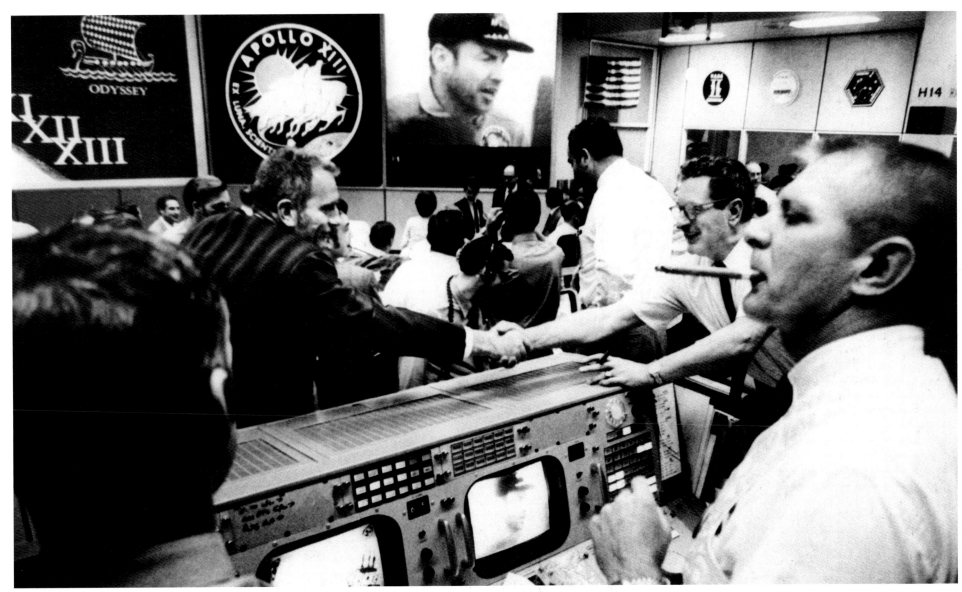

Looking like giant metallic insects, the lunar modules for *Apollo 14* (**opposite foreground**) and *Apollo 15* are given final checkout at the Kennedy Space Center in October 1970. By the time this picture was taken, NASA had cancelled the final three missions to the Moon to save funds.

Apollo 14 commander Alan Shepard (**above right**) and lunar module pilot Ed Mitchell train for their upcoming Moonwalks. The mission marked Shepard's return to spaceflight after a decade of being grounded due to an inner ear condition.

Alan Shepard (this page) lowers his gold-plated outer visor against the glare of the lunar sun shortly after taking his first steps onto the Moon on February 5, 1971.

The lunar module Antares (opposite) rests on the Moon's Fra Mauro highlands.

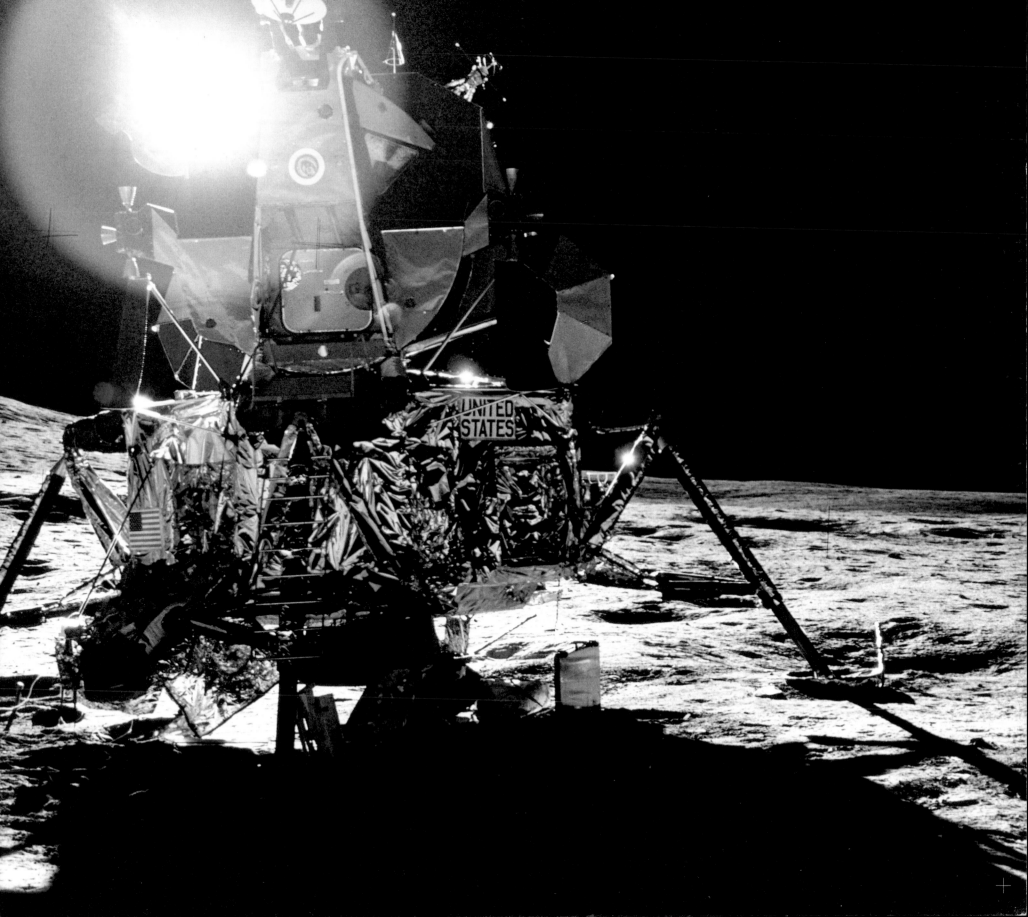

A gathering of a spaceflight fraternity (**left**) as astronauts host a pair of visiting cosmonauts at a party near the Houston space centre in 1970. From the left: Neil Armstrong, Buzz Aldrin, Bill Anders, Andrian Nikolayev, Jim McDivitt, Pete Conrad, Walt Cunningham, Tom Stafford, Jack Swigert, Dick Gordon, Rusty Schweickart, Dave Scott, Jim Lovell, Deke Slayton and Vitaly Sevastyanov.

By this time, the Soviet space programme had begun to shift its focus away from putting humans on the Moon to long-duration missions in Earth orbit. The panoramic view below is from the first automated lunar rover, *Lunakhod*, which roamed the Sea of Rains for almost a year beginning in November 1970.

The crew of the first Soviet space station, *Salyut 1*, who perished after spending 23 days in space when their craft developed a cabin air leak shortly before re-entry. **Clockwise from top left**: Viktor Patsayev, Georgy Dobrovolsky and Vladislav Volkov.

With the 15,000-foot (4,500 metre) summit of Mount Hadley looming behind him, Jim Irwin loads supplies onto the lunar rover during *Apollo 15*'s first Moonwalk on July 31, 1971.

Dave Scott at the rover with Hadley Rille beyond. Tiny dots visible on the floor of the winding, almost 1 mile (1.6 kilometre)-wide canyon are actually house-size boulders.

38

High on the slope of Mount Hadley Delta (**opposite**), Scott photographs a rock before collecting it during *Apollo 15*'s second Moonwalk on August 1, 1971. The following day, near the end of the mission's final excursion, Scott left a plaque (**above**) bearing the names of 14 deceased astronauts and cosmonauts, along with a small aluminium figure representing a fallen astronaut.

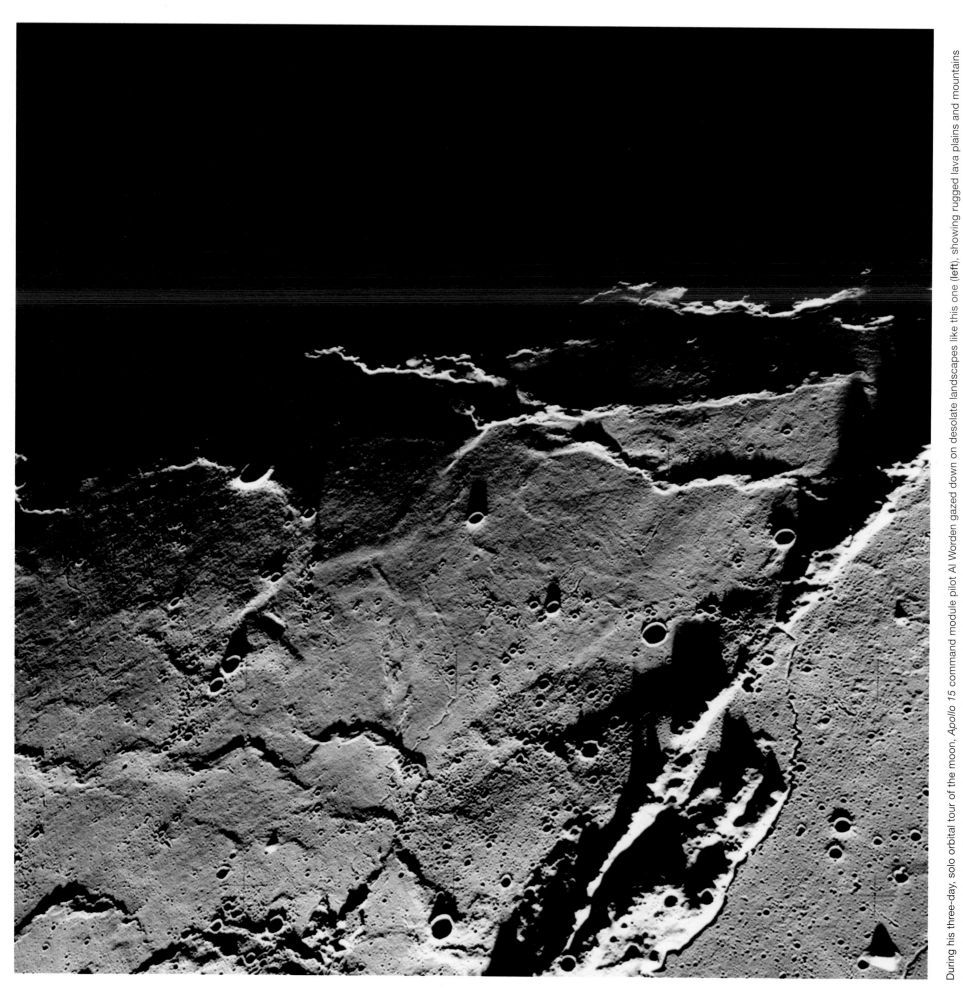

During his three-day, solo orbital tour of the moon, *Apollo 15* command module pilot Al Worden gazed down on desolate landscapes like this one (**left**), showing rugged lava plains and mountains near the crater Aristarchus. The *Apollo 15* crew had this view (**opposite**) of the Moon's Southern Sea as they began their homeward journey on August 4, 1971.

Near the end of the first *Apollo 16*'s first Moonwalk on April 21, 1972, mission commander John Young put the rover through its paces while Charlie Duke filmed the event. For all the excitement of the ride, this lunar Grand Prix was meant to provide data for engineers on how the rover handles in the Moon's one-sixth gravity.

Near the rim of giant North Ray crater, Duke works at the rover (**opposite**). Young approaches one of the more modest-sized boulders (**above**) he and Duke sampled during their third Moonwalk on April 23, 1972.

Apollo 17 lights up the sky (**right**) at the Kennedy Space Center shortly after midnight on December 7, 1972 as Gene Cernan, Ron Evans and Jack Schmitt begin the last lunar landing mission. In the first hours of their Moonward journey, the astronauts looked back on a full Earth (**opposite**), the only Apollo crew to see their planet fully illuminated. North is to the right of the image; the continent of Africa is visible, along with the Antarctic icecap on the left.

Gene Cernan test-driving the lunar rover (**above**) early in the first *Apollo 17* Moonwalk on December 11, 1972. With him during three days in the Taurus-Littrow valley is geologist-astronaut Jack Schmitt, seen with the American flag (**right**) and the distant Earth.

High on the slope of a mountain called the North Massif, Schmitt explores a house-size boulder the final *Apollo 17* Moonwalk, December 13, 1972. The excursion marked the last human footsteps on another world in the twentieth century.

Tired but elated at the success of *Apollo 17*'s explorations (**opposite**), Jack Schmitt smiles for Gene Cernan's camera inside the lunar module *Challenger*. Blackened with Moon dust, the helmets and spacesuits (**above**) used by the two explorers were stowed in the back of the tiny cabin.

Reunion, 69 miles (111 kilometres) above the Moon, December 14, 1972.
Ron Evans steers the command module *America* (**above**) toward a
docking with his returning crewmates. A battery of high-powered
cameras and scientific instruments is visible in the side of the service
module. As *Challenger* approaches (**below**), Cernan is visible in the
craft's left-hand window. Before leaving lunar orbit two days later, the
men photographed the Earthrise (**opposite**), a sight that has only
been witnessed by the Moon voyagers of the Apollo programme.

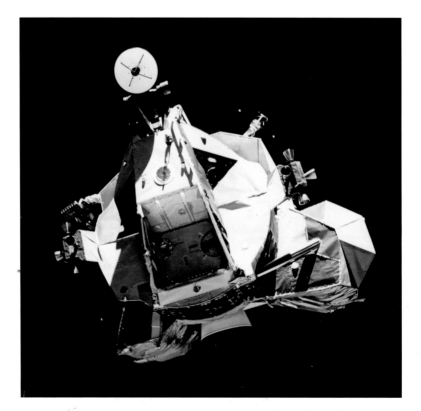

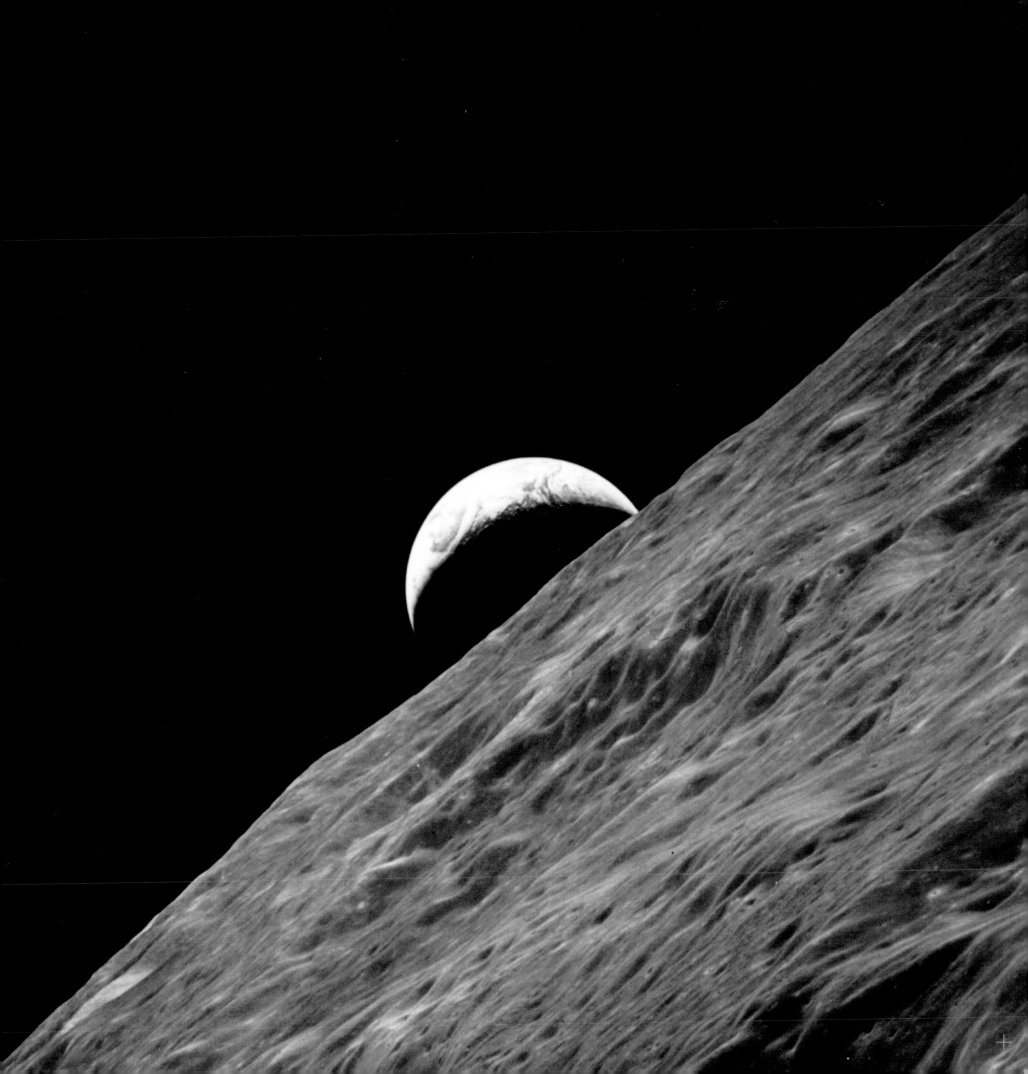

Floating in deep space, Ron Evans (**above**) retrieves canisters of scientific film from the service module during the homeward voyage. Jack Schmitt took this photograph while *Apollo 17* was approximately 180,000 miles (290,000 kilometres) from Earth, December 17, 1972. Just days later the Apollo programme came to an end as the command module *America* descended to the Pacific Ocean (**opposite**) under its three main parachutes.

The Soviet Union's Moon rocket, a giant N-1 booster, rests on its launch pad at the Baikonur Cosmodrome in Kazakhstan (**left**). In the third of four test-launches, this N-1 rose from the pad (**right**); seconds later, the rocket began to roll violently, then disintegrated. The fact that every N-1 launch failed effectively ruined the Soviets Union's chances of landing humans on the Moon. The mission called for a one-man lunar lander, designed to be flown almost entirely under automatic control (**opposite**).

The last *Saturn V* boosts the *Skylab* space station toward Earth orbit on May 14, 1973. Created from hardware culled from the Apollo programme, *Skylab* was home to three teams of astronauts who lived aboard the station for up to three months in 1973 and 1974. Because of a design flaw, the *Skylab*'s micrometeorite shield was torn off during launch, taking one of the station's twin solar panels with it. Spacewalking astronauts saved the station by installing a reflective sunshield on the hull (**opposite**) and freeing the remaining solar wing, which had been pinned to the station by a piece of debris.

With as much interior volume as a small house, *Skylab* offered ample room for its crews to conduct experiments, test new equipment and explore the delights of weightlessness. Joe Kerwin uses a straw to create a sphere of floating water during the first *Skylab* mission in June 1973 (**below**). Jack Lousma tries out the station's zero-gravity shower during the second mission (**opposite, top left**) in the summer of 1973; during the third mission in late 1973–early 1974, Gerry Carr tested a jet-powered backpack (**bottom left**) and "balanced" crewmate Bill Pogue on his fingertip

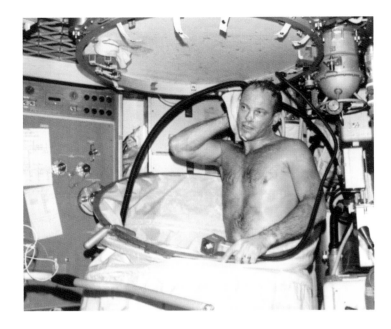

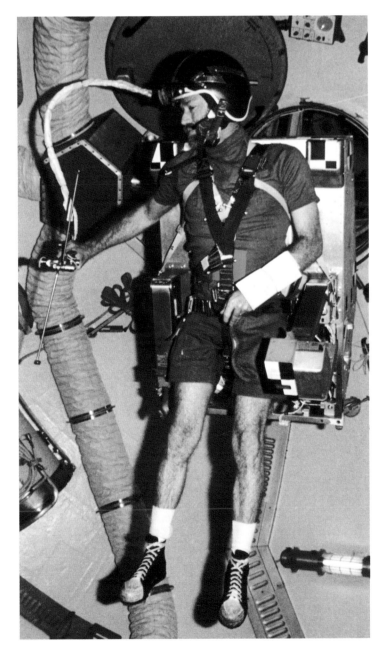

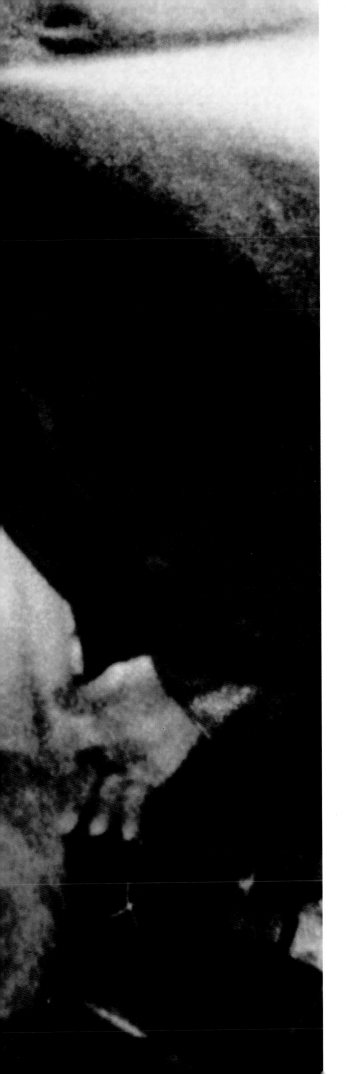

In the first international space docking on July 17, 1975 (**opposite**) Soyuz commander Alexei Leonov (**left**) shakes hands with Apollo commander Tom Stafford while Deke Slayton floats nearby in the U.S. docking module. Inside the Soyuz orbital module (**above**), Leonov entertains his American visitors.

The surface of Mercury, the closest planet to the Sun, had never been seen in detail before the U.S. *Mariner 10* probe photographed it in March 1974 (**opposite**). In three separate flybys in 1974–75, the craft sent back more than 3,500 images of a Moonlike, cratered world. The giant Caloris basin (**left**), an impact scar some 830 miles (1,335 kilometres) across, straddles the day–night boundary in a mosaic of Mariner images. After its formation, the floor of the vast crater was filled in by lava that erupted from Mercury's interior; since then, smaller craters have peppered the landscape. Averaging about 36 million miles (58 million kilometres) from the Sun, Mercury is 3,032 miles (4,879 kilometres) in diameter, about 40 per cent bigger than the Moon.

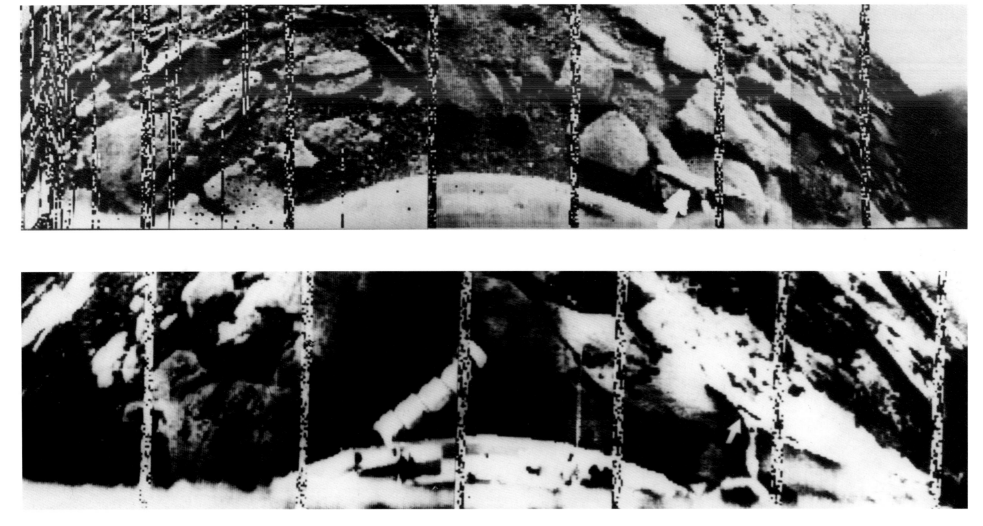

In October 1975 a pair of Soviet landers, *Venera 9* and *10*, achieved the extraordinary feat of landing on the surface of Venus and sending back the first images of its rock-strewn landscapes. Part of the spacecraft's support ring is visible at the bottom of each image; the "fisheye" appearance of the images is caused by the way the camera scanned the scene. Each lander lasted for about an hour before succumbing to the hellish surface conditions, which include temperatures of 850°Fahrenheit (454° Celsius) and pressures 90 times that on the surface of the Earth.

A Venera lander is checked out by technicians before launch.

A remarkably Earthlike scene photographed by the *Viking 1* lander on August 3, 1976, two weeks after touching down on the surface of Mars. Early morning sunlight (the local time is 7:30 a.m.) bathes rocks and drifts of fine sediment; the pair of large boulders at left – nicknamed "Big Joe" by mission scientists – is over 6 feet (1.8 metres) across and lies 26 feet (8 metres) from the camera. The white object bisecting the image is a boom tipped with meteorology sensors. The sky is bright due to suspended dust.

The *Viking 2* lander followed its predecessor to the Martian surface on September 3, 1976, and sent back these images of a rock-strewn desert. On the **left** is an image taken on May 18, 1979, which reveals a microscopic coating of frost, probably derived from water vapour from the thin Martian atmosphere that was carried to the site during seasonal dust storms. A month after landing, *Viking 2*'s surface sampler digs a trench (**below**) while collecting a scoop of Martian soil for analysis inside the spacecraft.

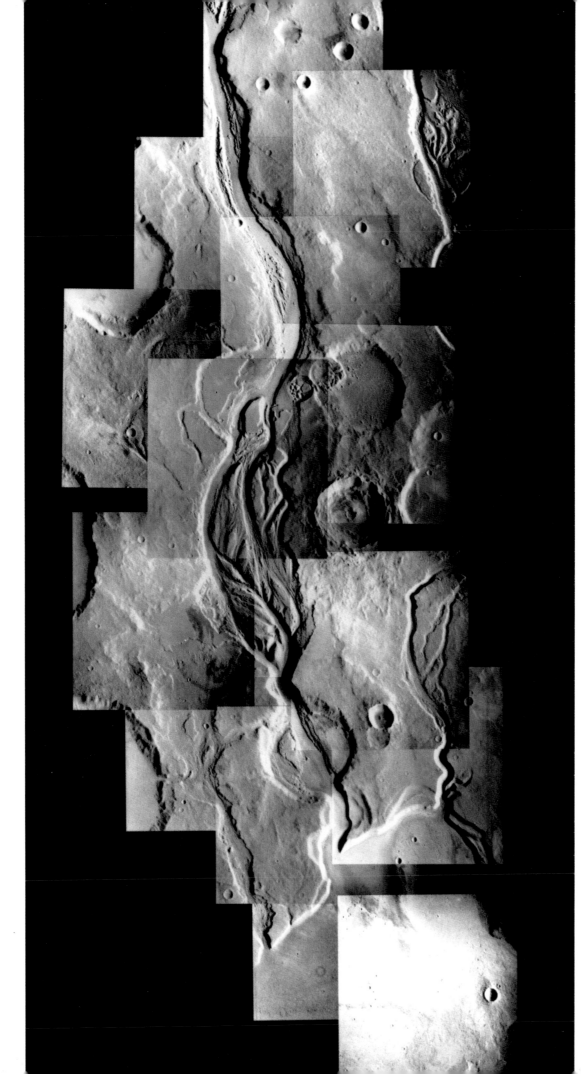

Spacecraft missions to Mars revealed that it is a world of geologic wonders, as shown in this mosaic of images from *Viking Orbiter 2* (**opposite**). A vast complex of canyons stretches across the equator, while four giant volcanoes are visible toward the planet's left edge. An ice cap of frozen water covers the north pole, and patches of light- and dark-coloured windblown dust stand out against the cratered plains. A network of channels – one of many discovered by orbiting spacecraft – attests to great floods that once scoured the surface of what is now a desert world (**left**).

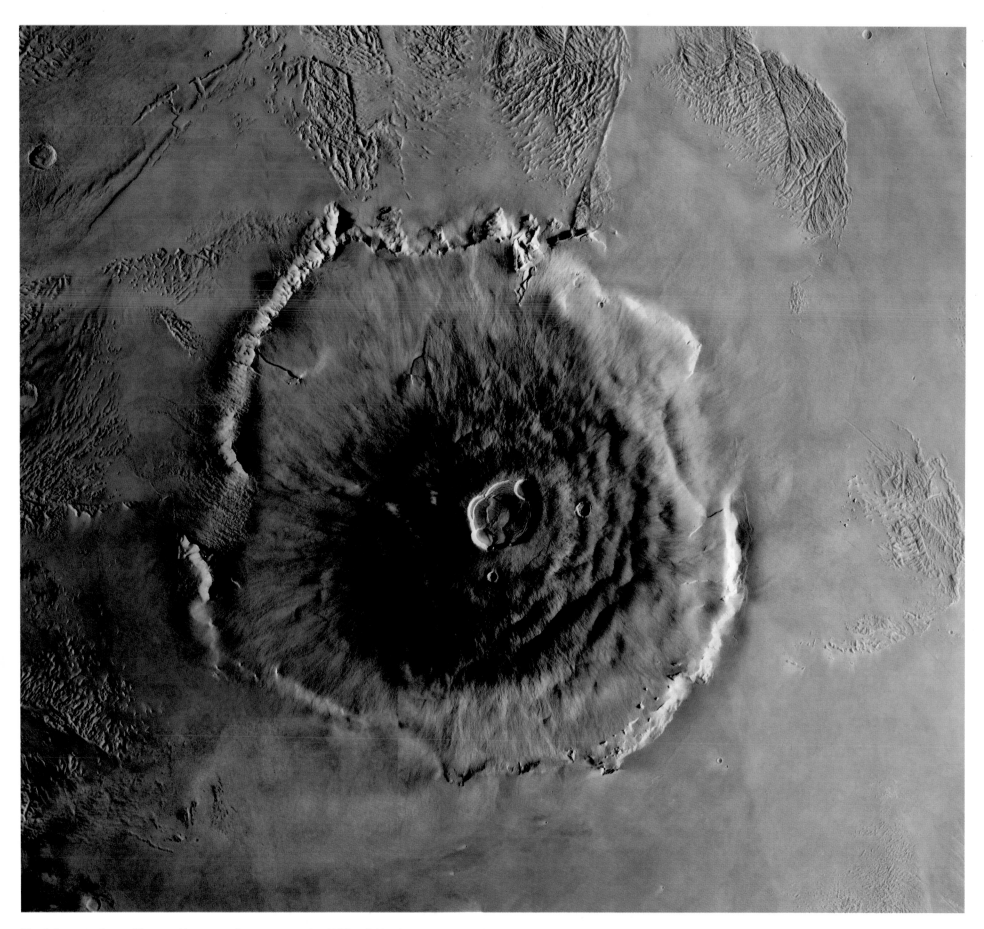

Mars's largest volcano, Olympus Mons, seen here as a mosaic of Viking Orbiter images, covers an area the size of Arizona and towers nearly 15 miles (24 kilometres) above the surrounding plains, almost three times the height of Mount Everest.

The summit caldera, or volcanic crater, seen in detail here is 56 miles (90 kilometres) across. The *Viking 1* orbiter obtained the images on these pages in 1977 and 1978.

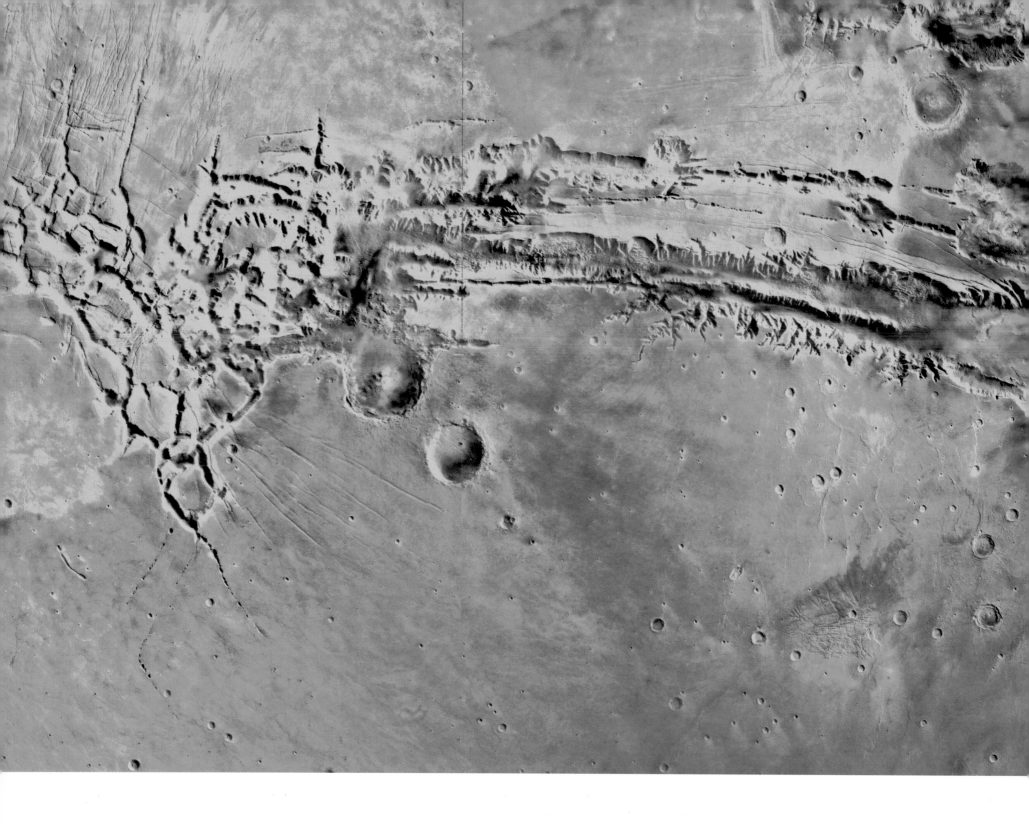

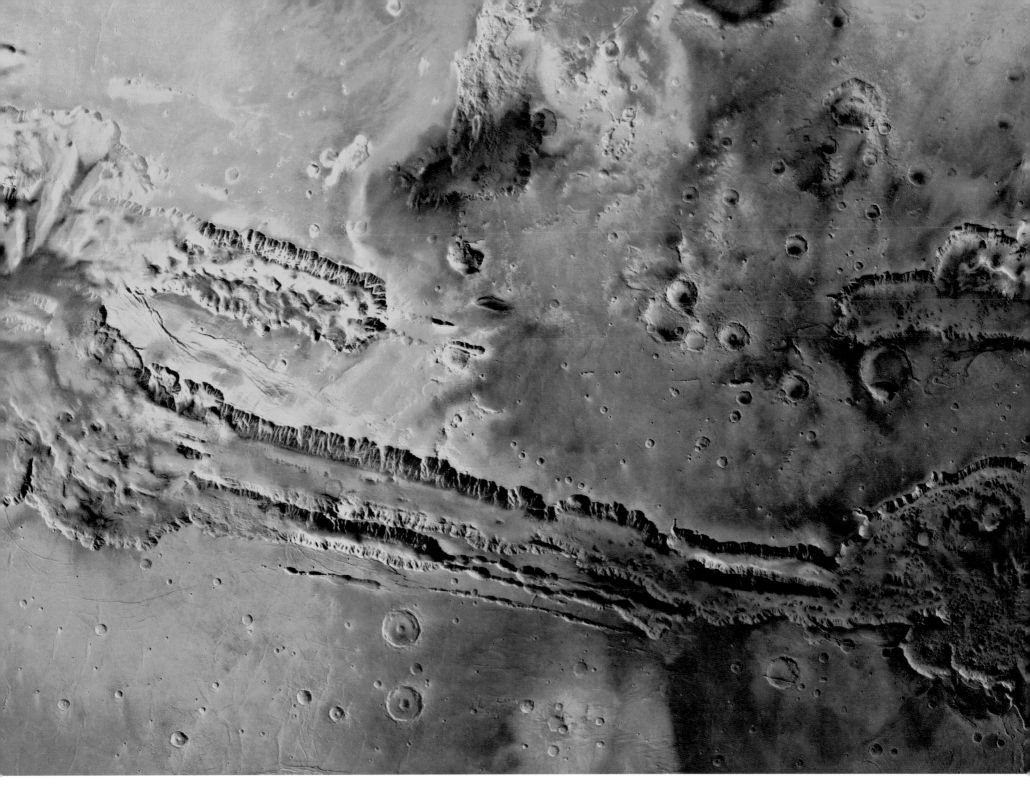

A mosaic of *Viking 1* orbiter images of the giant Martian canyon system, which was christened Valles Marineris after the *Mariner 9* orbiter that first photographed it in 1972. The canyons stretch more than 2,480 miles (3,991 kilometres) in length; some portions are more than 4 miles (6.5 kilometres) deep.

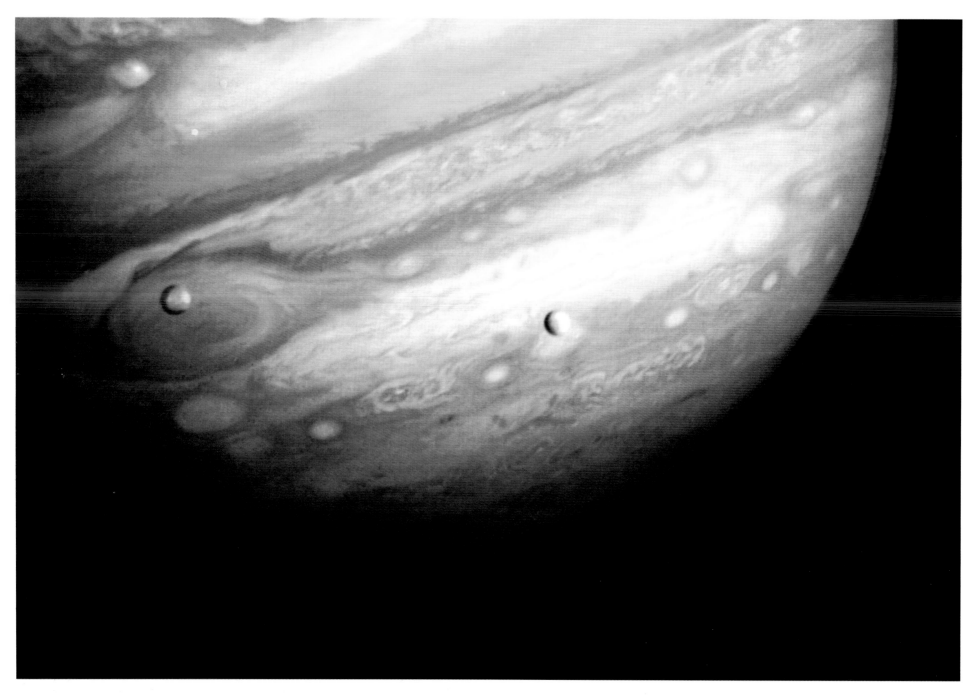

Jupiter, the solar system's largest planet, looms before the approaching *Voyager 1* spacecraft on February 23, 1979. Two of its larger moons, Io (**left**) and Europa, are seen against the planet's clouds. The Voyager flybys revealed stunning details in the turbulent and colourful atmosphere of Jupiter, such as the *Voyager 2* image (**opposite**) captured on July 3, 1979. The Great Red Spot, a storm twice as wide as Earth that has been observed for almost 400 years, dominates this view.

The Voyager flybys revealed that Jupiter's 2,200-mile (3,540 kilometre)-diameter moon Io is volcanically active, more so than any other body in the solar system. The discovery stemmed from a single image taken on March 8, 1979, as *Voyager 1* looked back on Io. The plume from one volcano arcs above the satellite's sunlit edge; the cloud from a second eruption stands out against the moon's night side. Further inspection of Voyager's high-resolution views of Io revealed dozens of eruptions, many of which contain molten sulphur. In the mosaic of images shown opposite, the volcano called Ra Patera (**upper right**) shows lava flows up to 200 miles (320 kilometres) long emanating from a dark central volcanic vent.

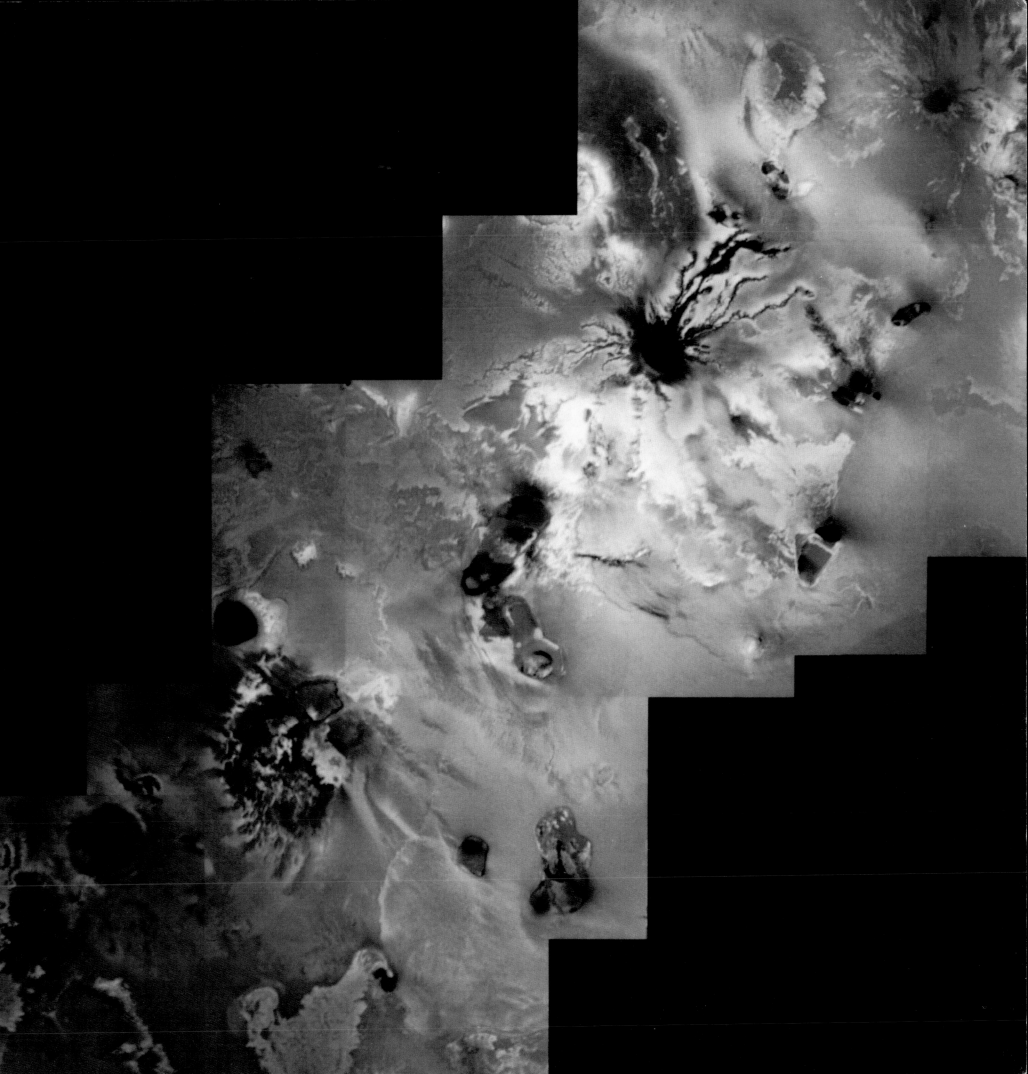

CHAPTER FOUR
SPACE SHUTTLES
AND SPACE STATIONS

On the morning of April 12, 1981, twenty years to the day after Yuri Gagarin became the world's first space traveller, a new kind of space vehicle stood ready for launch at the Kennedy Space Center in Florida. The Space Shuttle, the first reusable spacecraft, represented NASA's hopes for the future of space exploration. The winged shuttle orbiter, named *Columbia*, was the first in a planned fleet of shuttles, each of which was designed to be reused for as many as 100 missions. Supplying fuel to the orbiter's three liquid-fuelled main engines, which together delivered 4.3 million pounds (19 million newtons) of thrust, was a giant fuel tank mated to the orbiter. Two solid rocket boosters, attached to the external tank, would provide an additional 2.6 million pounds (11.5 million newtons) of thrust during launch.

With a 60-foot (18-metre)-long cargo bay, the orbiter was capable of carrying satellites and space probes and scientific and military payloads into orbit. It could even be used to retrieve an ailing satellite and bring it home. Unlike previous American spacecraft, the shuttle was designed to carry passengers – in this case, astronauts called mission specialists, who would do anything from conduct experiments to spacewalking. Also unlike its predecessors, the shuttle orbiter would land on a runway, and would be refurbished for its next flight, perhaps, NASA hoped, in as little as two weeks. The official name for the new craft, the Space Transportation System, reflected NASA's dream that the craft would replace other types of space vehicles and drastically reduce the cost of getting into Earth orbit. But the shuttle was also incredibly complex, and no one could be sure it would fly as designed. *Columbia*'s maiden voyage, officially designated STS-1, would reveal the answer. When *Columbia*'s engines ignited, spectators cheered the first American piloted launch in almost six years, while *Columbia*'s crew – mission commander John Young, a by the addition of new modules for various scientific purposes, offering its crews much more living and working space than the Salyuts. In December 1988 *Mir* was the scene of yet another record-breaking orbital voyage as Vladimir Titov and Musa Manarov completed a year-long mission. Ultimately, the Soviets hoped, the lessons learned from the space marathons would help make it possible to send humans to Mars.

For now, though, Mars itself went relatively unexplored. Instead, other worlds were the targets for historic encounters. Soviet and U.S. spacecraft equipped with radar-imaging systems to pierce the dense clouds of Venus gave scientists their first detailed look at the planet's surface features. In the outer solar system, *Voyager 1* and *2* flew past Saturn in 1981, exploring that planet's intricate rings and menagerie of moons. For *Voyager 1* that was the final encounter of the mission, but planners took advantage of the "Grand Tour" planetary alignment to send *Voyager 2* on to a flyby of Uranus in 1986 and a swing past Neptune in 1989. The Voyager probes, like *Pioneer 10* and *11*, were destined to leave the solar system, heading for the stars.

In 1986 Halley's comet, the most famous of the solar system's ancient, icy wanderers was surveyed by a small armada of robotic explorers during its passage around the Sun. Spacecraft from the Soviet Union gave tantalizing glimpses of the comet's solid portion, its nucleus. A pair of Japanese probes monitored Halley's halo of gas and dust, and NASA's International Cometary Explorer, already a veteran of history's first comet flyby in 1985, was targeted to fly through Halley's tail. But the star performer turned out to be Europe's *Giotto* probe, which passed close enough to Halley to snap the first detailed images ever made of a comet's nucleus. Those pictures revealed a very dark and irregularly shaped lump of ice and dust, measuring about 9 miles (14.5 kilometres) long and 5 miles (8 kilometres) wide. To some it may have resembled a cosmic potato, but to scientists, Halley's heart was a relic from the birth of the solar system.

The Halley missions were very visible evidence that the United States and the Soviet Union were not alone in the space business. China had launched its first satellite in 1970 and continued with satellites for military reconnaissance, communications, weather and scientific studies. China also developed its own family of powerful launch vehicles called Long March. Japan and India had joined the ranks of launching nations in the 1970s, while Israel launched its first satellite in 1988 and a vigorous and far-reaching space programme was spawned by Europe, where ten nations formed the European Space Agency (ESA) in 1975 (membership has since grown to 15 countries). Among ESA's achievements was the development of the Ariane series of boosters, first launched in 1979 from a specially built facility in French Guiana. Ariane's success as a commercial launch vehicle helped frustrate NASA's efforts to attract commercial customers for the Space Shuttle – but in the wake of the *Challenger* disaster, that role for the shuttle was discontinued.

As for the shuttle itself, there was good news in September 1988, when the orbiter *Discovery* logged a successful mission, putting NASA's piloted spaceflight programme back on track. A month later the Soviet Union launched their own shuttle orbiter from a new, powerful booster called *Energia*. The Soviet shuttle, called *Buran* (meaning snowstorm), closely resembled its American counterpart, although it used jet engines to assist in its runway landing. On its maiden voyage, a two-orbit mission, *Buran* was flown unpiloted, under automatic control. Although designed to launch and inspect satellites and to ferry cosmonauts to and from the *Mir* space station, *Buran* was essentially a military vehicle. Fearing that the American shuttle might threaten the Soviet Union's military space activities, they built *Buran* as a deterrent. However, budgetary problems killed the programme, and the craft's first flight was also its last. No one could predict, at that time, the drastic changes that lay just ahead for the Soviet Union and its space programme, or the unlikely turn of events that would unite the world's two superpowers, along with other nations, in space.

The shuttle *Discovery* lifts-off on November 22, 1989.

veteran of three previous space missions including the *Apollo 16* lunar landing, and rookie Bob Crippen – soared into orbit. After two days of checking *Columbia*'s systems, the pair prepared to return to Earth. That meant steering *Columbia* through a complex flight path, slowing from an orbital speed of 17,500 miles (28,000 kilometres) per hour to slightly more than 200 miles (320 kilometres) per hour at touchdown. To make matters more difficult, once the de-orbit rocket firing had been made the shuttle would be an unpowered glider, and the astronauts would have only one chance for the landing. Fortunately, Young and Crippen would be assisted by the craft's five on board computers. But there was another cause for concern, and it centred on *Columbia*'s thermal protection system.

To protect it from the fires of re-entry the orbiter was covered with thousands of tiles made from silica, but Young and Crippen could see that some of these tiles had been torn off by the stresses of launch. Whether any were missing from the craft's underside, which would bear the brunt of the re-entry heating, they could not determine. But images from classified tracking cameras on the ground told NASA that the orbiter would make it through re-entry without any trouble. On April 14 it did just that, and Young steered the craft to a flawless landing. In November *Columbia* returned to space, proving the shuttle's ability for re-use. After two more test missions, NASA declared the shuttle operational in July 1982. The shuttle era had begun.

During the next 3½ years three more orbiters joined *Columbia* in the shuttle fleet, and their crews demonstrated the remarkable capabilities of the Space Transportation System. By the end of 1985 NASA had racked up 24 flights that featured the first piloted missions to launch satellites, the first retrieval of satellites from orbit, a host of scientific experiments and a number of spacewalks. One of the more extraordinary moments came in 1984, when astronauts captured the ailing Solar Max astronomical satellite and, in a seven-hour spacewalk, repaired it before turning it loose. However, the shuttle had not achieved NASA's goal of reducing the cost of getting into orbit, in part because the maintenance of each craft between missions proved more extensive and more costly than anticipated. Even so, shuttle missions were flying at an unprecedented pace. In 1985 there had been nine shuttle flights, and NASA hoped to increase that number in 1986. It almost seemed that for all its dangers, spaceflight was becoming routine. NASA even approved a programme to let a handful of communicators – a schoolteacher, a journalist, and perhaps an artist – fly in space to share their experiences with the public.

All that changed on the morning of January 27, 1986, when the shuttle *Challenger* exploded 73 seconds after lift-off. On board were six astronauts and a schoolteacher from New Hampshire called Christa

McAuliffe. Mission controllers in Houston, along with spectators at the launch site and millions of television viewers around the world, could do nothing but watch in horror as *Challenger*'s crew perished. Months of investigations revealed that the disaster had been caused by a faulty seal within one of the craft's two solid rocket boosters – and that similar flaws had almost doomed earlier shuttle missions. These design errors would have to be corrected, and a host of other safety improvements made, before the shuttle could fly again. However, the optimism of the early shuttle program, suddenly erased by the *Challenger* accident, would be much harder to restore.

Even as the Space Shuttle saw triumph and tragedy, Soviet cosmonauts were pushing the limits of spaceflight endurance aboard the Salyut space stations. Salyut crews endured both physical and psychological stress in the course of their space marathons. The isolation of living and working in space for months on end was relieved only by occasional visits from Soyuz ferry crews, which included "guest cosmonauts" from nations like Hungary, Bulgaria and Cuba. Shipments of supplies from Earth in unpiloted Progress freighters contained fresh fruit, letters from home, and other treasures that were an antidote to boredom.

Above all, the marathon cosmonuats struggled to stay healthy in the face of the debilitating effects of long-term weightlessness, which causes muscles to atrophy, bones to lose minerals and the cardiovascula system to weaken. As a result, lengthy exercise sessions became an unpleasant but necessary part of the daily routine. In 1980, cosmonauts Leonid Popov and Valery Ryumin spent 185 days aboard *Salyut 6*, a record that would be eclipsed more than once in the years that followed. In 1984 a three-man crew remained in orbit for 237 days aboard *Salyut 7*. These missions gave new meaning to the term "long-duration spaceflight."

They also demonstrated the remarkable persistence and heroism of the cosmonauts, especially when Vladimir Dzhanibekov Viktor Savinykh rescued *Salyut 7*, which had lost contact with Earth after the departure of the previous crew. After arriving at Salyut in June 1985, Dzhanibekov and Savinykh realized its solar arrays had malfunctioned, starving the station of power. It was so cold inside the space station that frost covered the walls and equipment. Wearing winter clothing, including fur-lined hats, they worked for weeks to recharge batteries, thaw frozen water tanks, and make repairs. Weeks later, the station was back in working order and one of the most dramatic rescue operations in space history was complete.

The achievements of the Salyut programme were just a prelude to the Soviets' new space station, called *Mir* (peace), launched in 1986. The first modular space station, *Mir* was designed to be expanded

The first re-usable spacecraft, the Space Shuttle *Columbia*, heads for orbit on April 12, 1981 (**opposite**). At the controls are veteran John Young and rookie Bob Crippen, who went on to spend two days in space in the shuttle's maiden voyage. Two minutes into the launch (**left**) the twin solid rocket boosters, now spent, are jettisoned while the shuttle orbiter's three main engines continue to propel the craft into space; this view is from a high-powered tracking camera

Young steers *Columbia* to a landing at NASA's Dryden Flight Research Center (**above**) in California's Mojave desert on April 14. Joe Engle and Dick Truly (**opposite**) bring *Columbia* home from its second test-flight – proving the shuttle's ability for re-use – on November 14, 1981.

The seventh shuttle mission included the first American woman in space, scientist-astronaut Sally Ride, seen on the flight deck of the orbiter *Challenger* (**opposite**). During the mission the astronauts released a free-flying satellite called SPAS that photographed *Challenger* against a backdrop of clouds (**above**).

A self-contained satellite, Bruce McCandless tests the Manned Manoeuvring Unit, a backpack powered by jets of nitrogen gas. McCandless and crewmate Bob Stewart flew the MMU up to 300 feet (91 metres) from *Challenger* on February 7, 1984.

Shuttle mission STS-51A featured the retrieval of two errant satellites by astronauts using the Manned Manoeuvring Unit. Outfitted with a special docking device, Dale Gardner (**opposite**) approaches the Westar VI satellite, November 14, 1984. With the two satellites safely in the cargo bay of the shuttle *Discovery* (**above**), Gardner poses for fellow spacewalker Joe Allen. The sign he holds indicates what will happen to the satellites after they are returned to Earth.

In history's worst spaceflight disaster, the seven-member crew of the shuttle *Challenger* perished when the craft exploded 73 seconds after liftoff on January 28, 1986. **Above, from left:** *Challenger*'s crew included schoolteacher Christa McAuliffe, payload specialist Greg Jarvis, mission specialist Judy Resnik, mission commander Dick Scobee, mission specialist Ron McNair, pilot Mike Smith and mission specialist Ellison Onizuka. The two solid rocket boosters leave twisted contrails as they career away from the exploding shuttle (**opposite**).

Preceding pages. An image from *Voyager 1* as it approached Saturn and its moons Dione (left-hand page) and Tethys on November 3, 1980. The shadows of Tethys and Saturn's three bright rings appear on the planet's cloudtops. In a *Voyager 2* close-up (right-hand page, top left) taken on August 23, 1981, Saturn is dimly visible through its rings, which are made of countless particles of ice and dust. The icy Saturnian moon Mimas (top right), 248 miles across (400 kilometres), is scarred by the 80-mile- (130-kilometre)-wide Herschel crater that gives it a resemblance to the Death Star from the movie *Star Wars*. Four days after its closest approach to Saturn, *Voyager 1* looked back and photographed the planet as a crescent (bottom), a view never seen from Earth.

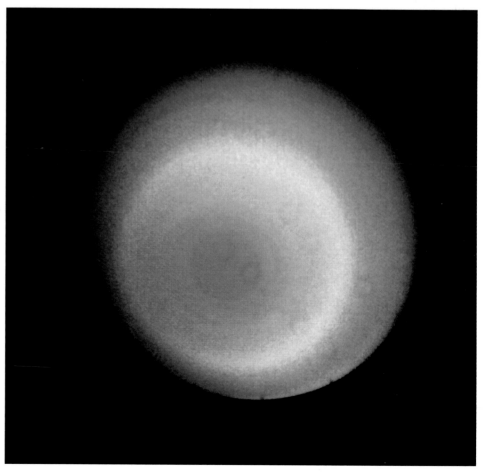

The Solar System's seventh planet, Uranus, was the target of *Voyager 2*'s flyby on January 24, 1986. In natural light, the planet is a featureless, pale blue ball (top right); processing in false-colour reveals features within the planet's methane-rich atmosphere (bottom right). One of the strangest worlds visited by *Voyager*, the Uranian moon Miranda (opposite) is 280 miles (450 kilometres) in diameter and thought to be composed of a mixture of ice and rock. The jumbled surface of the moon suggests that volcanic activity began to alter Miranda billions of years ago, but ceased before the transformation could be completed.

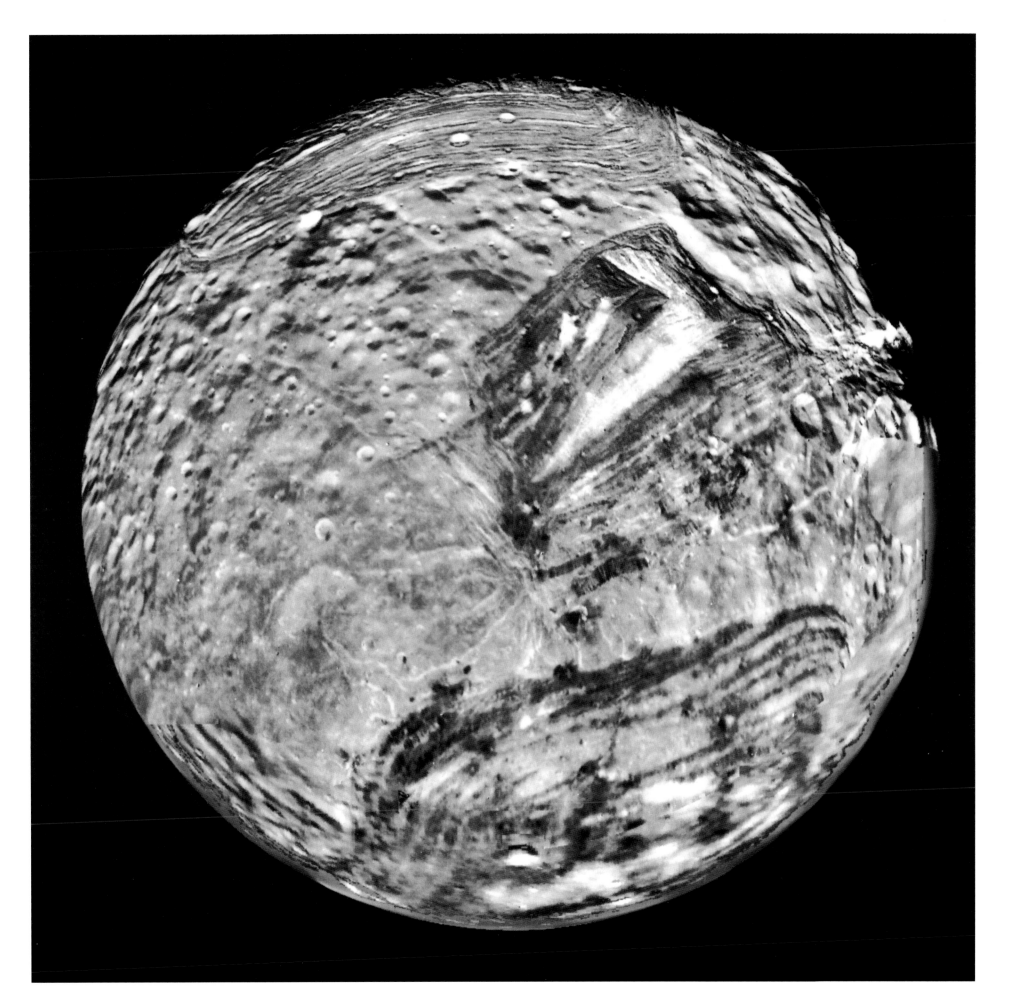

The nucleus of Halley's Comet, potato-shaped, 10 miles (16 kilometres) long and black as coal, spews jets of gas and dust into space (**above**). The European probe *Giotto* passed within 370 miles (595 kilometres) of the famous icy wanderer on March 13, 1986.

Twelve years after launch and 2.8 billion miles (4.5 billion kilometres) from the Sun, *Voyager 2* swept past Neptune on August 24-25 (**opposite**), 1989. Neptune, which is almost four times larger than Earth, has an atmosphere rich in methane, which helps create the planet's blueish colour. At the time of *Voyager 2*'s encounter a dark, Earth-sized storm raged in Neptune's southern hemisphere.

The *Salyut 7* space station (**left**) was the scene of marathon Soviet space missions in the 1980s. In 1984 a trio of cosmonauts spent 8 months aboard the station. In the Soyuz simulator at Star City (**top**), the cosmonauts' village near Moscow, Valery Ryumin (**left**) and Leonid Popov train for a six-month mission to *Salyut 6* in 1980. Remarkably, Ryumin had already logged half a year aboard the station in 1979 when he volunteered for the new marathon. One feature of the Salyut missions was participation by cosmonauts from other nations. Czech pilot Vladimir Remek (**above, centre**) and Soviet cosmonaut Alexei Gubarev (**right**) made a week-long visit to *Salyut 6* in March, 1978, where Georgy Grechko (**left**) and Yuri Romanenko were near the end of a 96-day mission.

Mission commander Vladimir Kovalenok (**opposite left**) and flight engineer Alexandr Ivanchenkov take time out for a zero-gravity game of chess during a 136-day residence aboard *Salyut 6* in 1978. The mission, nearly two months longer than the final U.S. *Skylab* visit, gave the Soviets a clear lead in experience with long-duration spaceflight.

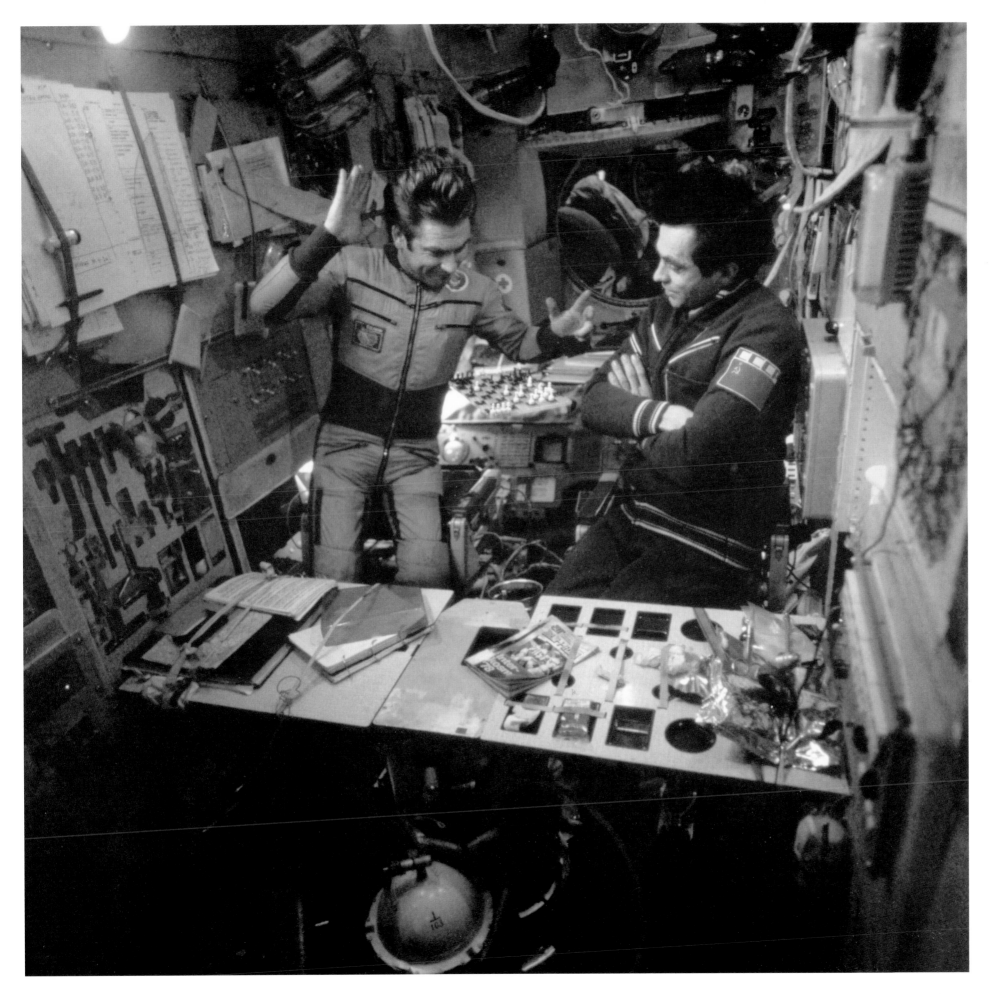

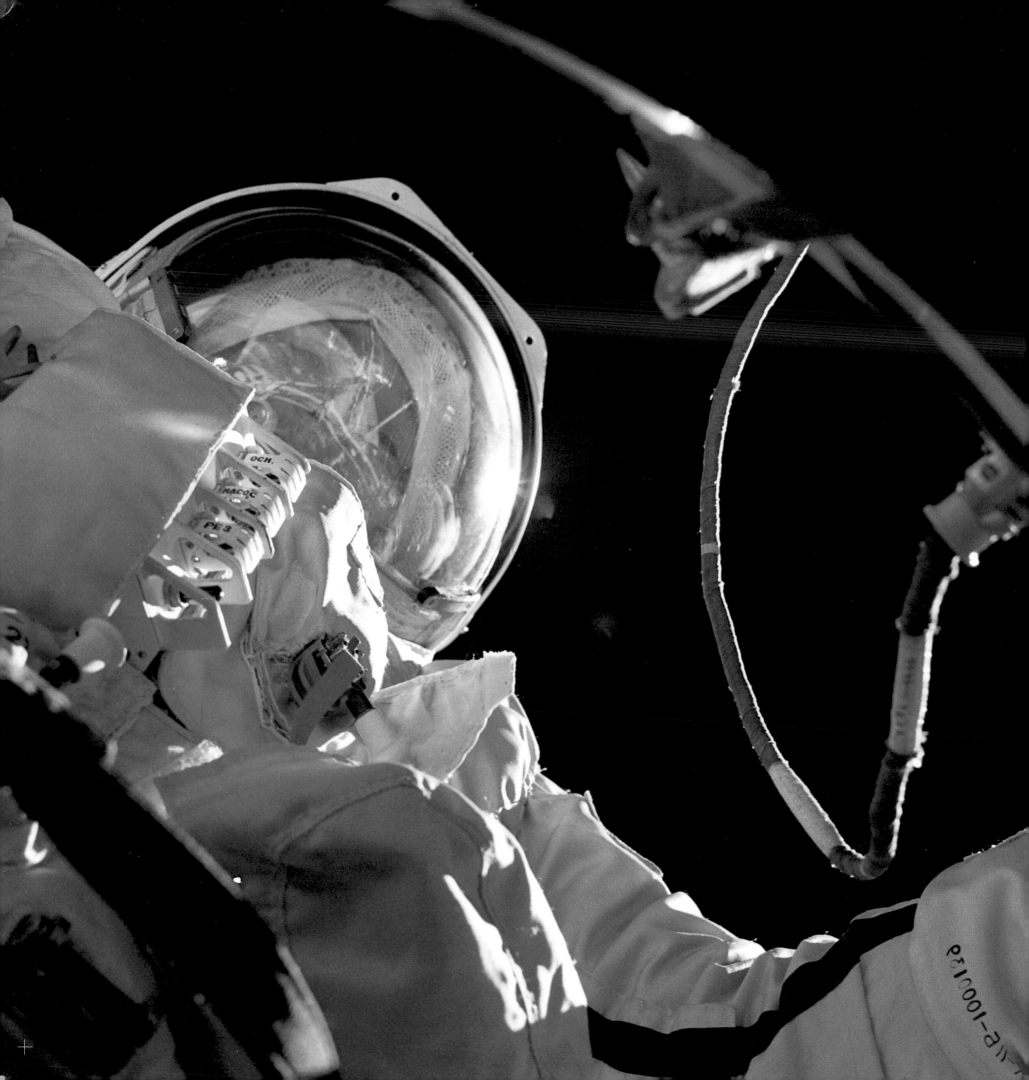

In the diary of his 211-day mission to *Salyut 7*, Valentin Lebedev (**opposite**) wrote of the nervousness he felt before making a spacewalk. Here he is seen on his first trip outside, July 30, 1982, when he and Anatoly Berezovoy tested methods for space construction projects.

The first modular space station, *Mir*, in various stages of completion (**right**). The station's core module (**top**) was launched in February 1986, with additional modules being added over the next 10 years. During his six months aboard *Mir* in 1987, Alexandr Alexandrov (**below**) enjoys a watermelon, part of a newly arrived shipment from Earth aboard a Progress supply freighter. Such amenities helped ease the burden of isolation during the long and difficult months of the space marathon.

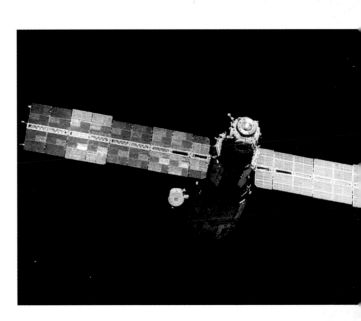

The Soviet Union's space shuttle, *Buran* ("snowstorm") bore a strong resemblance to its U.S. counterpart, but was designed to be flown automatically. It is seen mated to the *Energia* heavy-lift booster (**opposite, far left**), being raised into position on the launch pad at Baikonur (**opposite, left**) and during cross-country transport on top of the giant An-225 Mriya. *Buran* made a single, unpiloted test flight in November 1988 before the programme was cancelled.

The European Space Agency's Ariane family of boosters, first launched in 1979, has become a mainstay of commercial space activities. Shown here is the *Ariane 4* launcher; the ESA launch site is located in Kourou in French Guiana.

CHAPTER FIVE
THROUGH RUGGED WAYS
TO THE STARS

In 1990, anyone who needed a reminder that exploration demands persistence could find it in astronomer Lyman Spitzer, Jr. In 1946, Spitzer proposed an idea that must have seemed fantastic to his colleagues: a telescope launched into orbit that would observe the heavens from space, free of the blurring and distortions caused by Earth's turbulent atmosphere. Such a telescope, Spitzer said, might alter our fundamental understanding of the universe. It took more than four decades for Spitzer's dream to be realized, but on April 24, 1990, the Space Shuttle *Discovery* lifted off carrying the Hubble Space Telescope (HST). Spitzer, then aged 75, was at the Florida launch site to witness the event. A day later HST (named for pioneering astronomer Edwin Hubble, who discovered the expansion of the universe) was released by *Discovery*'s crew into orbit 388 miles (620 kilometres) above the Earth. Astronomers eagerly awaited the telescope's first views of the heavens.

When those images came, however, they were disappointingly blurry. Something was wrong with the telescope's primary mirror, which had cost more than $500 million to produce. Analysts soon determined that in its final manufacture the mirror had been ground slightly too flat at the edges. The error was minuscule – invisible to the human eye – but that was enough to prevent Hubble from achieving the image resolution expected. Astronomers who had been awaiting Hubble's launch, delayed four years after the *Challenger* accident, were devastated. NASA came under stinging criticism for botching the $1.5-billion project. But soon a plan emerged to save the telescope, with the aid of Space Shuttle astronauts; fortunately, HST was designed to be serviced in space by shuttle crews. Experts designed a set of special lenses that, once installed inside the telescope, would function as a pair of hi-tech eyeglasses to correct the problem. The repair work would require the most delicate and demanding spacewalks ever made.

In December 1993, the shuttle *Endeavour* soared into a night sky above Florida, carrying the Hubble repair crew. After reaching the telescope and placing it in *Endeavour*'s cargo bay using the orbiter's robotic arm, the astronauts conducted a record five spacewalks to install the corrective optics inside the telescope. Then they returned HST to its orbital perch, knowing that for all their hard work, the repair might not be successful. But fears vanished when Hubble's new images came down: the view was superb, ten times more detailed than images from ground-based telescopes. Even before the repair, astronomers had managed to make scientific gains with the telescope, but now, they were ecstatic: HST performed even better than expected, opening a new era of astronomical discovery in space. With it astronomers could conduct a stellar census in distant star clusters and glean new clues to the mysterious workings of black holes. Hubble zeroed in on the hearts of distant galaxies and peered into stellar nurseries in our own Milky Way galaxy. In 1994 it recorded the impact of Comet Shoemaker Levy 9 into the atmosphere of Jupiter, a spectacular demonstration of the force that has done the most to shape the solar system since its birth. Hubble's rescue also repaired NASA's tarnished image; the mission is still considered by many to be the Space Shuttle's finest hour.

However, there were other problems weighing on NASA in 1993 and they centred on the next big project in space: a permanent, inhabited space station in Earth orbit. The Reagan administration had approved the station in 1984 but it subsequently became mired in bureaucracy. Now, with not a single piece of hardware launched, the project was seriously behind schedule and over budget. The station, which had already been redesigned more than once to save funds, was now in danger of being cancelled. Ironically, it was rescued by the end of the Cold War that had done so much to spur previous space efforts.

Following the dissolution of the Soviet Union in 1991, the Russian space programme, victim of a tattered economy, was on the brink of extinction. In the West there were concerns that Russian engineers and scientists once engaged in peaceful pursuits might turn their energies to making weapons. NASA, searching for antidotes to the station's rising costs, had already explored the idea of using Russia's planned *Mir-2* as a core station module and the Soyuz ferry as an emergency escape vehicle for station crews. So in 1993 Bill Clinton's administration directed NASA to make Russia a full partner in the station project. Renamed the International Space Station (ISS), the effort would bring together 16 nations, including Canada, Japan and the countries of the European Space Agency. When completed, it would weigh a million pounds (2.2 million kilogrammes), contain six research laboratories and take up as much area as a football field. NASA hoped the ISS, with a crew of as many as seven astronauts and visiting scientists, would make discoveries that justified its multi-billion-dollar cost.

Before that could happen, however, there was the tricky business of integrating the human spaceflight programmes of the two former Cold War rivals. The focus of this effort became the *Mir* space station, still occupied nearly a decade after its 1986 launch. Plans emerged to have cosmonauts fly on the Space Shuttle; in addition, American astronauts would visit *Mir* for several months at a time, working under Russian crews. In March 1995, physician-astronaut Norm Thagard became the first American to ride a Soyuz ferry into space, beginning an orbital tour of 115 days. In June, Thagard's ride home, the shuttle *Atlantis*, became the first American. spacecraft to dock with *Mir* – picking up where the Apollo-Soyuz mission left off some 20 years earlier. Thagard's marathon got the United States back in the business of long-duration spaceflight for the first time since the *Skylab* missions of the early 1970s. But there was no question about who the space endurance leaders were: Thagard's visit to *Mir* overlapped with a stay of 14 months by cosmonaut Valeri Polyakov, a record that still stands.

For more than three years the visits of Americans to *Mir* continued and while these missions included some notable successes – for example, Shannon Lucid's 188-day flight, which set a U.S. space endurance record yet unbroken – they also included some of the most harrowing moments in spaceflight history. In February 1997 a fire broke out aboard *Mir* when an oxygen-generating "candle" malfunctioned, spewing a two-foot (0.3 metre) flame and sending thick, acrid smoke through the station for several long minutes. Nearly forced to abandon ship, *Mir*'s crew, including astronaut Jerry Linenger, managed to hang on long enough for the fire to burn itself out. The emergency, which caused no lasting effects, was followed by more difficulties, including leaks of toxic coolant inside the station. On Earth, friction intensified between managers at the Russian Space Agency and their NASA counterparts, who did not always see eye-to-eye about the conduct of the missions.

In space, the troubles aboard *Mir* worsened. In June 1997 an unmanned Progress cargo freighter collided with the station during a docking attempt, puncturing one module and damaging a solar panel. With the station losing pressure, cosmonauts Vasily Tsibliyev and Aleksandr Lazutkin and astronaut Mike Foale struggled to seal off the damaged module. Then came the long battle of restoring power to the station, now dark, silent and slowly tumbling. As the crisis continued, so did criticism of the station in the United States, where *Mir* was seen as an orbiting "white elephant". But thanks largely to the remarkable courage and persistence of its crews, *Mir* recovered from its difficulties and for all its problems – in space hardships allowed only one planetary mission in the 1990s, a Mars orbiter with a collection of small landers, but the craft never made it out of Earth orbit. In 1993 NASA suffered the failure of its billion-dollar Mars Observer probe, which was lost just as it was attempting to go into orbit around Mars; analysts later determined that the craft had probably suffered a massive fuel leak. The failure came as NASA faced continuing budget cuts from Congress, and it forced the agency to take a new approach to exploration of the solar system. Dubbed "faster, better, cheaper" it focused on smaller, less expensive probes that would also take less time to develop. For many scientists this was a welcome change following the 1980s when, amid tightening budgets, only two American missions in the solar system – the *Magellan* Venus radar mapping mission and the *Galileo* Jupiter orbiter – had been launched.

The challenge for NASA, then, became doing more with less, and "faster, better, cheaper" got off to a good start. In 1997 Mars *Pathfinder* became the first spacecraft since *Viking 1* and *2* to land on the Red Planet. With a new type of landing system that employed airbags instead of complex retro-rockets, *Pathfinder* "bumped down" onto a rock-strewn valley on July 4. A day later the craft released a diminutive rover called *Sojourner*, which prowled for weeks among the boulders under control from Earth. The mission generated enormous public interest, thanks in part to the Internet, which allowed computer users to view brand new images of Mars soon after they were received on Earth. September saw the arrival of the Mars Global Surveyor orbiter, carrying the high-resolution camera originally slated for the Mars Observer mission; by November it was taking the most detailed views of the planet yet obtained, and making discoveries that would ultimately rewrite scientists' understanding of the Red Planet.

These achievements seemed to herald a new era of Martian exploration, culminating with a planned sample return mission early in the twenty-first century. But success was soon followed by disappointment: in December 1999 NASA lost four Mars probes, including the first lander slated to explore the planet's polar regions. The failures, later attributed to simple mistakes by mission planners and engineers, raised questions about whether the risks of "faster, better, cheaper" outweighed the gains, nevertheless, everyone knew there was no going back to the high-cost, big-ticket missions of years past. They also knew the scientific stakes for continuing were higher than ever, because they now included a renewed search for life beyond our home planet. The prospects for past life on Mars, for so long the target of speculation, got a boost when NASA scientists announced they had found possible evidence for fossil bacteria inside a Martian meteorite. Although the finding was extremely controversial – and remains so today – it helped move the search for extraterrestrial life from the fringes of space science to the mainstream. Then, another world joined Mars as a possible abode for living things: Jupiter's icy moon Europa, which fell under the scrutiny of the *Galileo* orbiter. *Galileo*'s images helped demonstrate that an ocean of liquid water lies beneath Europa's frozen crust, fuelling speculation that such an ocean might harbour organisms. The possibility of solving these mysteries was a lure for the planners of missions to come, in the twenty-first century.

With such exciting prospects for unmanned exploration, NASA's human spaceflight programme seemed by comparison to be lacking vision. Even the space station programme seemed in need of some higher purpose, something to rekindle the sense of mission felt during the heady days of the Cold War. As if to amplify such nostalgia, a Cold Warrior returned to space in October 1998: John Glenn, the first American in orbit, spent nine days in orbit aboard the shuttle *Discovery*. At the age of 77, Glenn – who by this time had been a U.S. Senator for 25 years – became the world's oldest space traveller. His *Discovery* crewmates had been children at the time of his Mercury mission; growing up with the space programme had helped inspire them to become astronauts. Glenn's participation in the flight focused on the effects of aging on human response to spaceflight; NASA hoped the data might one day lead to advances in the understanding of aging on Earth.

In the meantime, Glenn's return to space gave the shuttle programme an enormous boost in public interest. In a sense, it heralded the end of one era and the beginning of another; by year's end the first two modules of the International Space Station were in orbit, beginning an assembly process that would continue into the first years of the new century. History's largest construction project would pose considerable challenges, not the least of which being that it would take place 200 miles (320 kilometres) above the Earth; and if it succeeded, greater challenges lay ahead: to find uses for the station that would change the world for the better.

The newly repaired Hubble Space Telescope floats on the end of the shuttle *Endeavour*'s robotic manipulator arm (**opposite**) shortly before being released to resume its explorations of the heavens on December 9, 1993.

A gallery of images from the Hubble Space Telescope begins
with scenes relatively close to home.

Hubble captured this view of the solar system's most distant world, Pluto
(**above**) and its satellite Charon on February 21, 1994. Dark blotches in the
atmosphere of Jupiter (**right**) mark the impact of fragments of the comet
Shoemaker-Levy 9 in July 1994. A dying star 2,100 light years from Earth
(**opposite**) spews twin jets of glowing gas, into space. Astronomers have
measured the velocity of the jets at 200 miles (320 kilometres) per second.

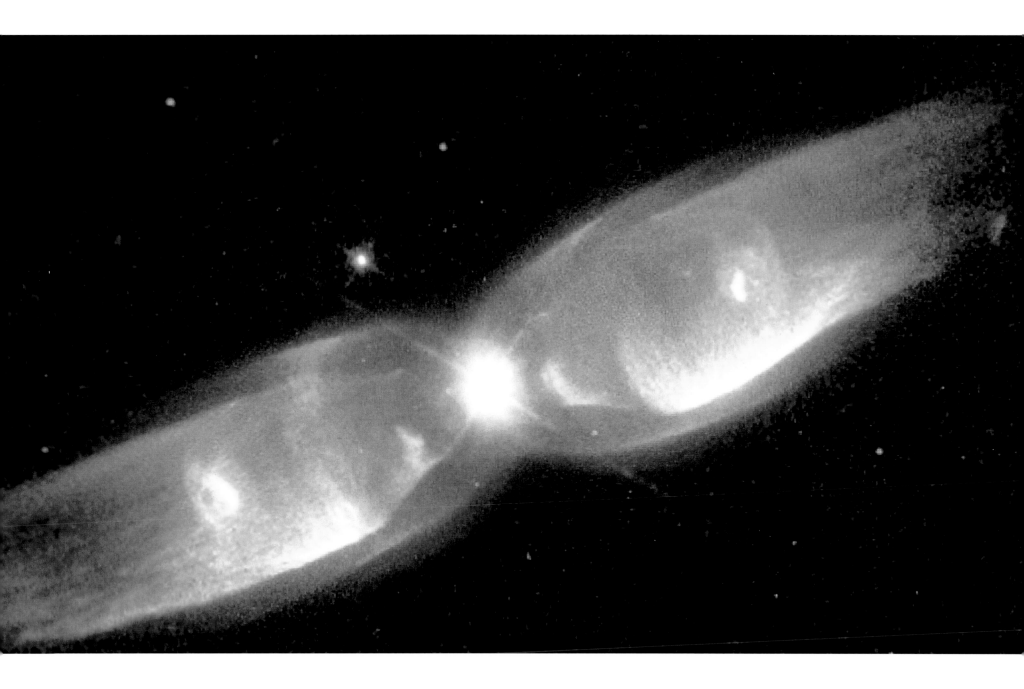

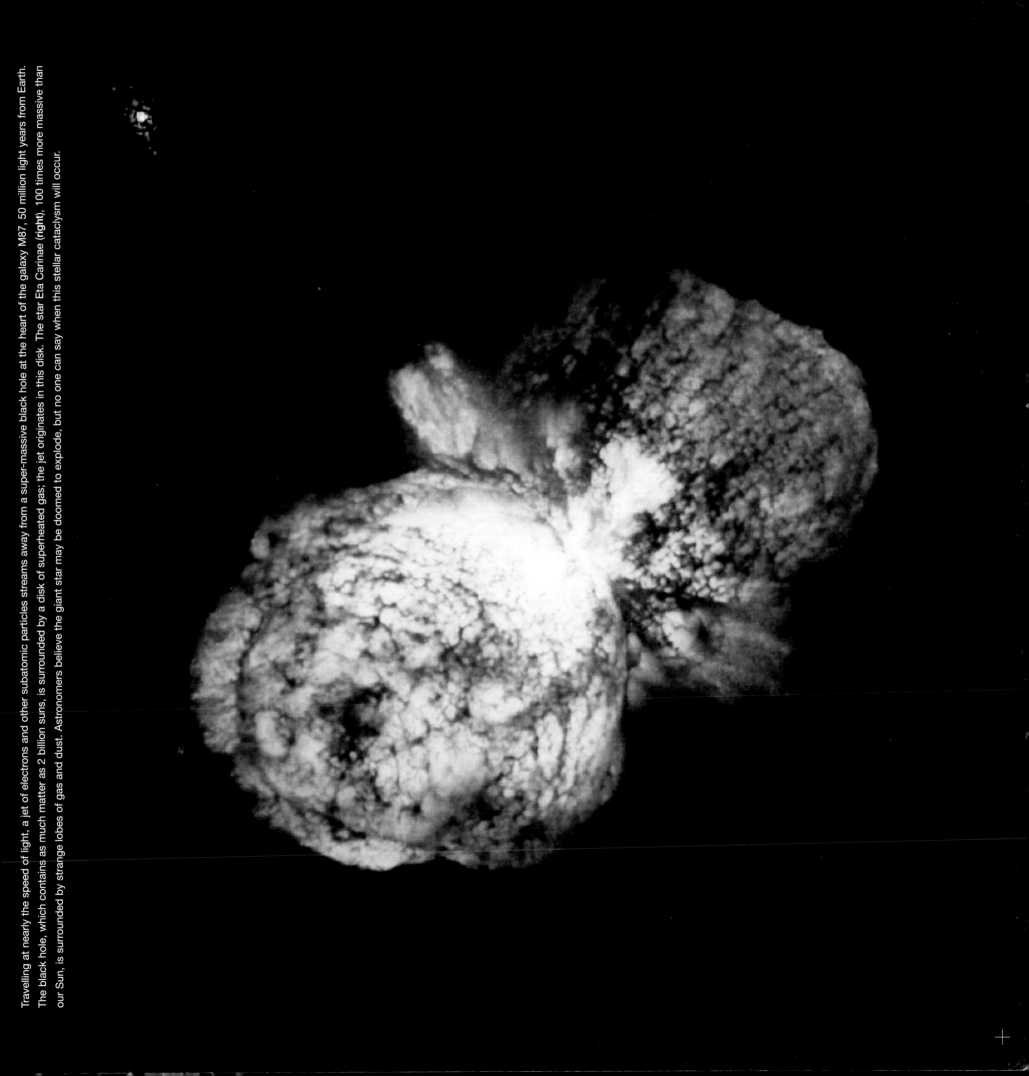

Travelling at nearly the speed of light, a jet of electrons and other subatomic particles streams away from a super-massive black hole at the heart of the galaxy M87, 50 million light years from Earth. The black hole, which contains as much matter as 2 billion suns, is surrounded by a disk of superheated gas; the jet originates in this disk. The star Eta Carinae (**right**), 100 times more massive than our Sun, is surrounded by strange lobes of gas and dust. Astronomers believe the giant star may be doomed to explode, but no one can say when this stellar cataclysm will occur.

Two spiral galaxies (**above**), trapped in orbit around each other 114 million light years from Earth; the smaller of the two is about the size of our Milky Way galaxy. Some 2 billion light years from Earth lies a cluster of galaxies (**below**) whose gravitational pull is so powerful that it bends light; the arc-shaped patterns visible in this picture are actually the distorted images of more distant galaxies.

Looking back to the earliest epochs of cosmic history, Hubble zeroed in on a tiny patch of sky in the constellation Ursa Major and revealed a glittering array of some 1,500 extremely distant galaxies. These galaxies, each of which is four billion times fainter than can be seen by the human eye, had never been seen by the largest telescopes on Earth. This picture gives astronomers a look back to a time when the universe was perhaps less than a billion years old, yielding valuable clues to how galaxies evolved.

The space shuttle *Columbia* leaves Earth on June 25, 1992 carrying the first U.S. Microgravity Laboratory (**opposite**), for a two-week mission devoted to experiments in space manufacturing.

Standing on special supports in the cargo bay of the space shuttle *Endeavour* (**above**), astronauts Pierre Thuot, Rick Hieb and Tom Akers hold onto the errant Intelsat VI-3, which they repaired and released in to Earth orbit. The astronauts' previous efforts to capture the satellite had been unsuccessful, necessitating the unprecedented three-man spacewalk on May 13, 1992.

Thermal protection tiles and blankets, which shield the shuttle from the intense heat of re-entry, cover the orbiter *Discovery* (**left**) as photographed during the STS-92 mission.

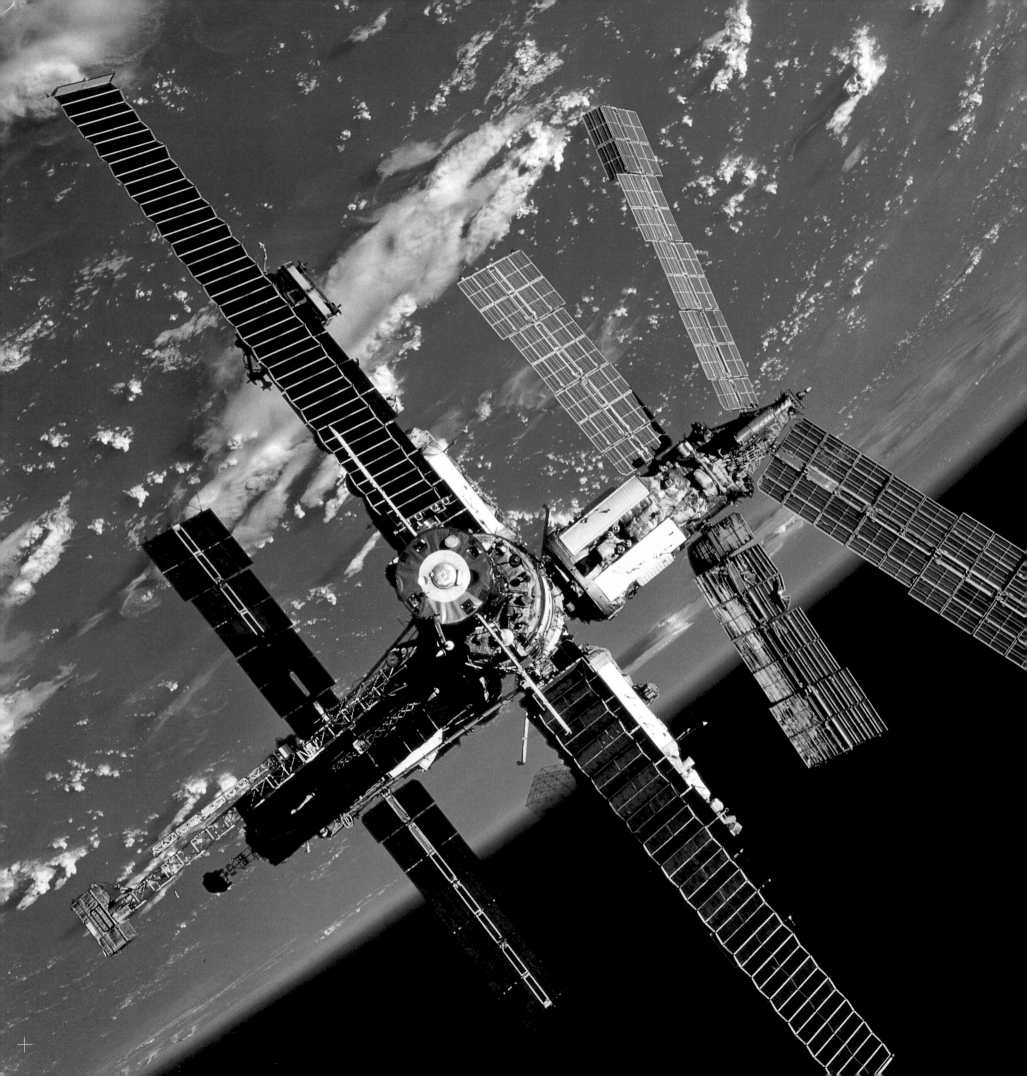

The missions of the space shuttle to the Russian *Mir* space station marked a new era in space cooperation. STS-71 commander Hoot Gibson (**above, right**) is greeted by *Mir-18* commander Vladimir Dezhurov on June 29, 1995. By 1997, eleven years after the launch of the station's core module, *Mir*'s construction had been completed with six modules, solar power arrays and a docking compartment totalling 100 tons (101.6 tonnes).

Staying in space for months at a time takes a toll on the human body that must be combated with regular exercise sessions. During her six-month visit to *Mir* Shannon Lucid (**above right**) is seen using one of two treadmills aboard the station.

Mir-23 flight engineer Aleksandr Lazutkin (**right**) works with a medical experiment to study the effects of zero-gravity on his cardiovascular system.

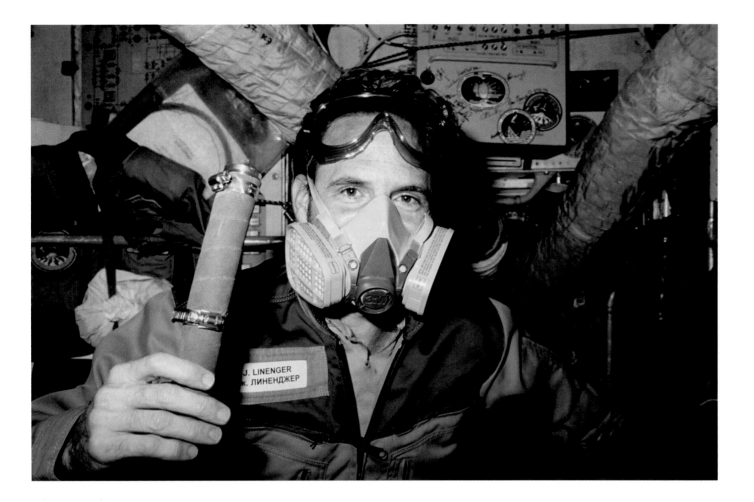

The U.S. visits to *Mir* were marked by a seemingly endless string of problems aboard the ageing station, including some of the worst crises in spaceflight history. Corroded coolant pipes leaked antifreeze and had to be repaired, as Jerry Linenger (**left**) demonstrated during his 132-day mission in 1997. Linenger was on board on February 23, 1997 when a serious fire broke out aboard *Mir* due to a malfunctioning oxygen generator, filling the station with thick smoke. On June 25, 1997, during Mike Foale's 134-day flight, an unmanned Progress cargo freighter struck *Mir*, damaging a solar panel (**below**) and punching a hole in the station's *Spektr* module. Foale and his Russian crewmates had to work quickly to seal off the leak.

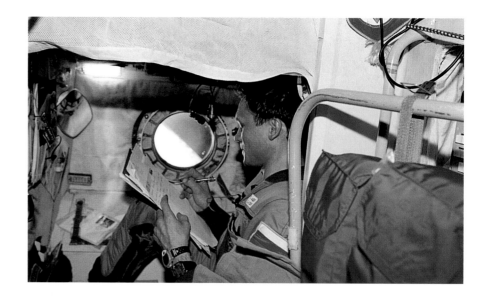

Leisure time aboard *Mir* was scarce, but the station's residents had no shortage of diversions. *Mir-20* flight engineer Sergei Avdeyev writes a letter in his sleep compartment (**top**). Gennady Strekalov of the *Mir-18* crew plays a guitar in the *Mir* core module (**above**).

The view of Earth from 240 miles (386 kilometres) was always spectacular; this *Mir-22* photo (**opposite**) shows an orbital Moonrise over central Asia.

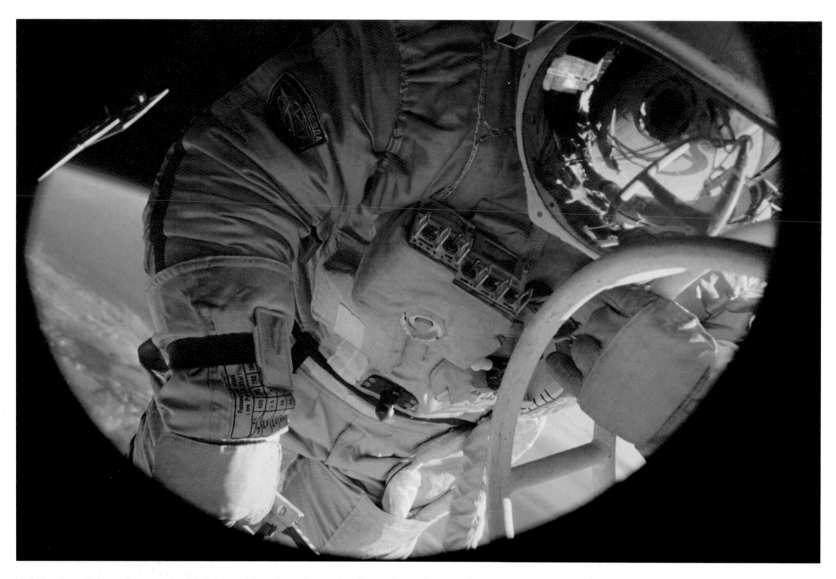

Yuri Usachev, flight engineer on the *Mir-21* crew (**above**), works outside the station during one of five spacewalks he made during his 193-day mission in 1996. Usachev had logged a previous six-month stay aboard the station two years earlier.

Carrying the returning *Mir-18* cosmonauts and American guest-cosmonaut Norm Thagard, the shuttle *Atlantis* (**opposite**) backs away from *Mir* on July 4, 1995. Thagard, the first American to live aboard *Mir*, spent nearly four months aboard the station. This photo was taken from a free-flying Soyuz spacecraft carrying cosmonauts Anatoliy Solovyev and Nikolai Budarin, the *Mir-19* crew.

Site of an ancient Martian flood, Ares Vallis (Mars Valley) was the landing spot for the Mars Pathfinder mission in 1997. The probe deployed a small rover called *Sojourner* (**far left)**, that roamed among the rocks making chemical analyses and sending back pictures that included the Pathfinder lander (**left**, with the airbags used to land) and Martian boulders (**bottom**). Pathfinder also sent back views of the sun setting over the rugged horizon (**below**).

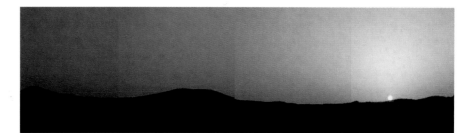

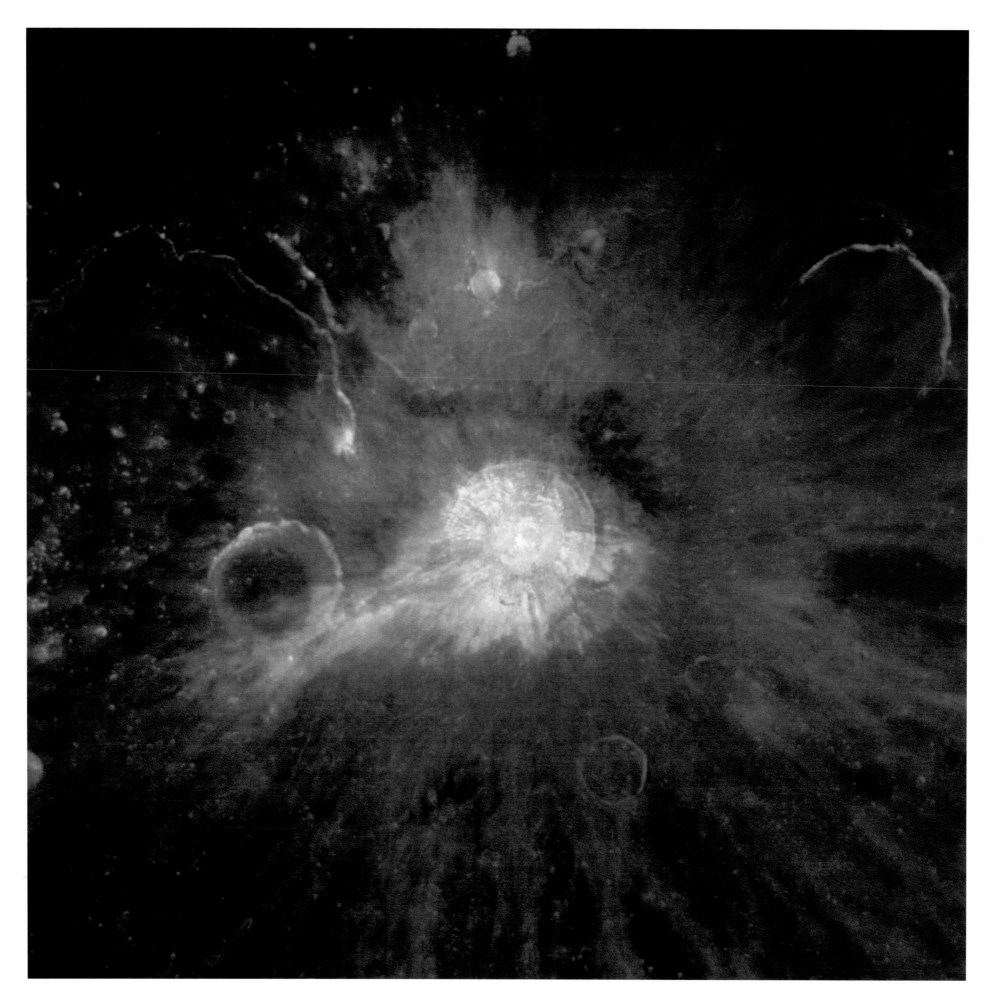

For the first time, scientists obtained a global map of the Moon's composition from a U.S. spacecraft called *Clementine*, which made an orbital survey 1994. The crater Aristarchus (**opposite**) and its surrounding plateau are among the most geologically diverse regions on the Moon. Venus shines brightly to the left of the Moon (**above**) in this view from *Clementine*'s star-tracker camera.

Overleaf. A tortured landscape of ice, the surface of Jupiter's moon Europa is seen from NASA's *Galileo* orbiter, which began a reconnaissance of the giant planet and its family of satellites in late 1995. In this false-colour mosaic, which shows features as small as 750 feet (229 metres) across, a network of ridges (yellow and brown hues) that probably formed by eruptions of fresh ice crisscross an older, smoother surface (blueish hues). *Galileo*'s images have provided strong evidence that an ocean of liquid water lies beneath Europa's icy crust. Finding out whether that ocean harbours life will undoubtedly be a goal of missions later in the twenty-first century.

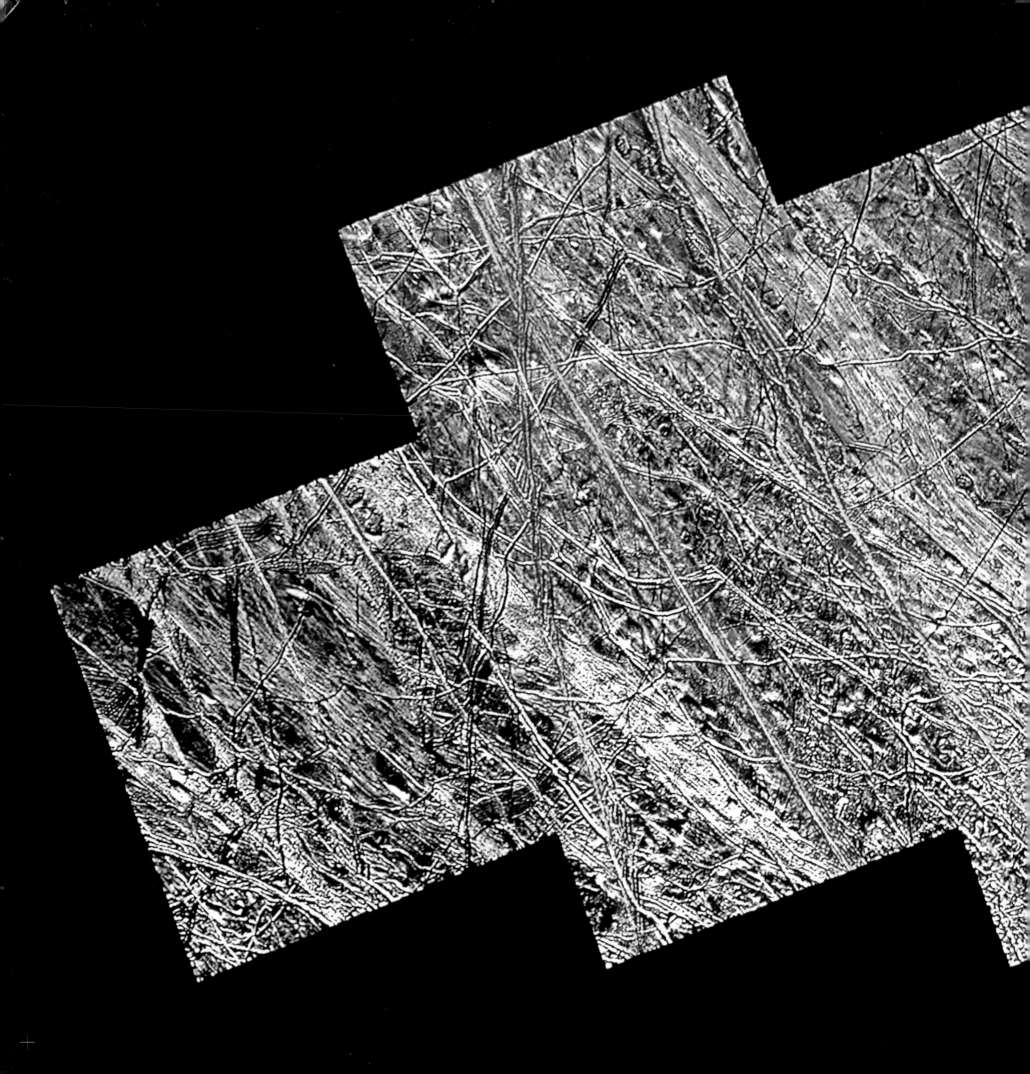

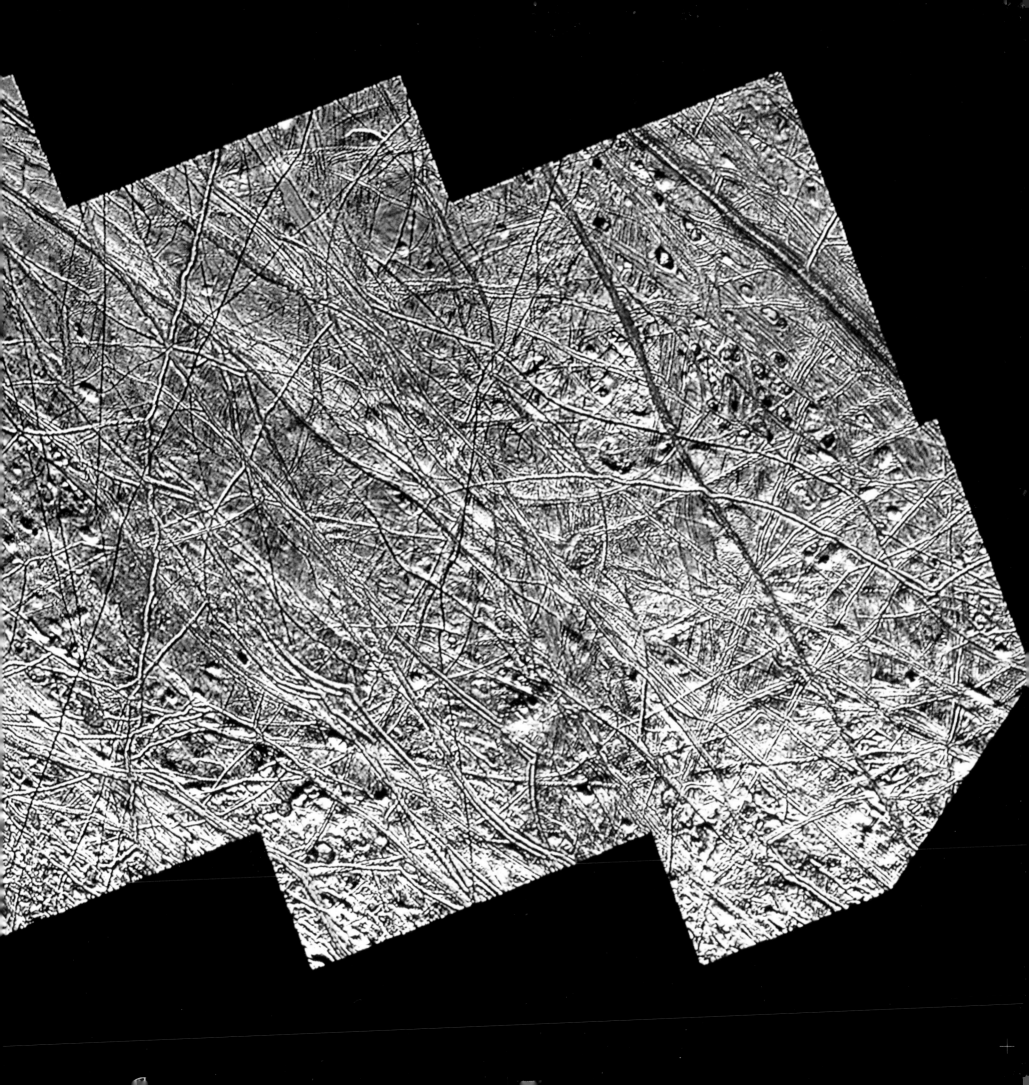

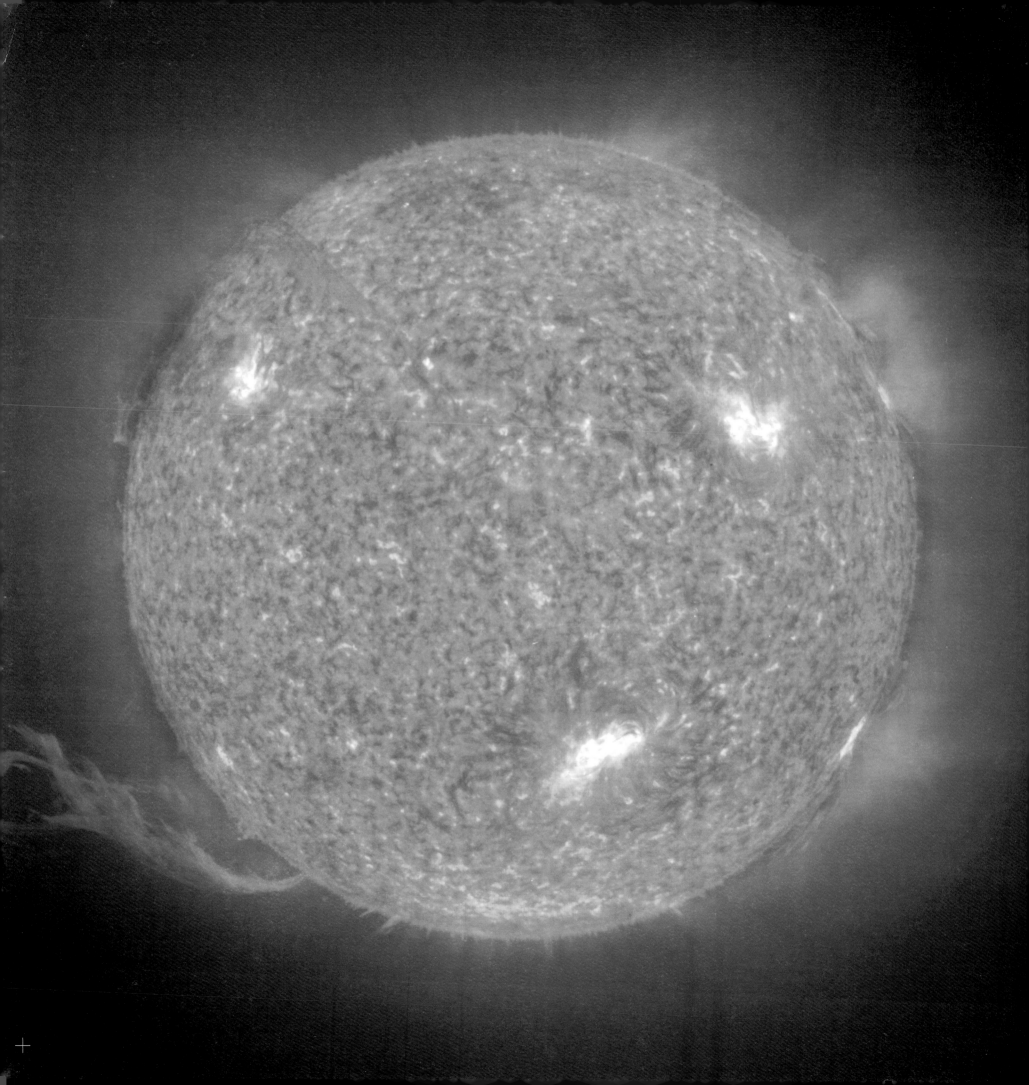

Keeping a constant vigil on the Sun's violent activity, the Solar and Heliospheric Observatory (SOHO) spacecraft, a joint project of the European Space Agency and NASA, was launched in December 1995. On September 14, 1997, SOHO took this image in ultraviolet light showing an enormous eruption of gas, called a solar prominence, streaming into space. The Sun's intense magnetic field heats the gases of its outer atmosphere, or corona, to millions of degrees, a phenomenon studied by the Transition Region and Coronal Explorer (TRACE) satellite. TRACE captured this view of glowing gases following loops of magnetic energy on January 10, 2001.

Still a space pioneer at the age of 77, John Glenn poses with his crewmates aboard the shuttle *Discovery* during his return to space in October 1998. **Clockwise** from Glenn: mission commander Curt Brown, pilot Steve Lindsey and mission specialists Steve Robinson, Pedro Duque of Italy and Chiaki Mukai of Japan.

In one of three spacewalks during a December 1999 repair mission to the Hubble Space Telescope, Steve Smith retrieves a power tool while anchored at the end of the shuttle *Discovery*'s robotic arm. *Discovery*'s astronauts replaced the telescopes 10-year-old batteries, installed a new, advanced computer, and made other modifications to prepare the Hubble for its second decade of exploring the heavens.

CHAPTER SIX
THE TWENTY-FIRST CENTURY

If you happened to be in just the right spot on March 23, 2001, you had the chance to witness the passing of a space legend when the *Mir* space station, under command from mission control near Moscow, fell to Earth in a shower of brilliant meteors over the South Pacific. There were people, especially in the former Soviet Union, who would have liked to see the 15-year-old veteran kept in service somehow. But many others, including NASA managers, urgently wanted the already strained resources of the Russian manned space programme focused on the daunting tasks of building and operating the International Space Station.

Even as *Mir* met its fiery end, the ISS era was underway. In November 2000 the station's first expedition crew, American astronaut Bill Shepherd and Russian cosmonauts Sergei Krikalev and Yuri Gidzenko, had arrived at the station, which they christened *Alpha*, for a 4½-month tour. In early March 2001 they were replaced by the Expedition Two crew in a pattern NASA hoped to sustain in a continuous occupation until at least the year 2016. With the addition of the U.S. *Destiny* laboratory module, *Alpha* was poised to begin a programme of scientific research that NASA hoped would justify the time, money, and effort required to create the station.

By mid-2001, however, it was clear that wasn't going to be easy. NASA revealed that the ISS programme was suffering from a budget shortfall of several billion dollars (the exact figure was not certain), threatening the station's completion, which was slated for 2006. The problems with the ISS only underscored the enormous and ongoing challenges of opening the space frontier – challenges that were both technological and economic. Four decades after the first humans reached Earth orbit, it still cost about 10 thousand dollars per pound to put them there. For years, space advocates had hoped that some yet undiscovered benefit of space research would offset this cost by sparking a kind of gold rush in Earth orbit. Some predicted the lure would be manufacturing pharmaceuticals or other valuable products in zero gravity. There was even talk of space tourism, especially in April 2001, when the crew of a Soyuz taxi mission to the ISS included California businessman Dennis Tito, who paid the Russians a reported 20 million dollars to spend 10 days in Earth orbit. So far, however, no space-age gold rush is in sight.

Meanwhile, there was better news from robotic missions. A spacecraft named *Mars Odyssey* entered martian orbit in October 2001 and soon began mapping surface minerals and searching for water, paving the way for possible future missions to search for life on Mars. A Saturn orbiter called *Cassini* was well on its way, carrying a European-built probe called *Huygens* that will land on the planet's mysterious moon, Titan in November 2004. In March 2002 shuttle astronauts installed a new camera on the Hubble Space Telescope, extending HST's already prodigious reach into the heavens. Joining Hubble in space, along with the *Chandra* x-ray observatory, was NASA's new infrared telescope named for astronomer Lyman Spitzer, launched in August 2003. On January 2, 2004 a probe called *Stardust* completed its rendezvous with Comet Wild 2, collecting samples of its ancient dust for return to Earth in 2006. Also in January 2004, NASA's two new Mars rovers, named *Spirit* and *Opportunity*, landed successfully and began their explorations as robotic field geologists. In a thrilling discovery, their instruments revealed the first "ground-truth" evidence that liquid water had once flowed on Mars.

There were even some unplanned achievements. The NEAR-Shoemaker spacecraft completed its one-year orbital reconnaissance of the asteroid Eros so successfully that mission planners decided to try a bold experiment, ordering the craft to land on the asteroid's surface – and on February 12, 2001, they succeeded. The mission gave scientists a closer look at the kind of interplanetary wanderer that might one day pose a threat to life on our own world, like the asteroid that wiped out the dinosaurs some 65 million years ago. As science fiction writer Larry Niven has pointed out, "The dinosaurs became extinct because they didn't have a space programme." With that in mind, it isn't hard to view space exploration as a means of ensuring human survival on this planet.

But when it came to exploring space with people, the difficulties only worsened. On February 1, 2003, after a 16-day science mission, the Space Shuttle *Columbia* disintegrated during reentry, killing its seven-member crew. For a second time, tragedy brought the shuttle programme to a halt, and forced NASA into a painful period of self-examination. As with the *Challenger* disaster 17 years before, the accident could be traced partly to a flawed understanding of the risks involved in flying the shuttle. From now on, there could be no doubt that the Space Shuttle remained an experimental vehicle. With Russia's Soyuz rockets now the only means of sending new crews to the International Space Station, NASA faced not only the daunting task of recovering from the *Columbia* tragedy, but renewed uncertainty about the shuttle's future, and the future of human space flight. Only for China, who launched its first astronaut aboard the *Shenzhou 5* spacecraft on October 15, 2003, was there the excitement of new beginnings in space.

Then, in January 2004, U.S. President George W. Bush called on NASA to send humans back to the moon by 2020, and then, to the surface of Mars, all in the name of scientific discovery. With the new space initiative came sweeping changes: the shuttle would be retired in 2010, after completion of the ISS and NASA would develop a Crew Exploration Vehicle to carry astronauts beyond Earth orbit for the first time since the *Apollo* moon missions. It was clear, however, that carrying out the new directive would be as difficult as any goal in NASA's history. Even aside from the technical and medical challenges of sending humans on long deep-space voyages – the hazards of space radiation, for example, are daunting – there was the political hurdle of convincing the American public and its leaders that such a programme is worth its projected multi-billion-dollar cost. At this writing, the fate of the Bush space initiative is uncertain.

Still, no one doubts that our goals in space can become reality – if we choose to make them so. In that we can take inspiration from one of our own creations, the *Pioneer 10* spacecraft. Launched in 1972 on the first mission to Jupiter, and with a design lifetime of 21 months, *Pioneer 10* lasted decades. In 1983 it became the first manmade object to cross the orbit of Pluto. And on March 1, 2002, 30 years after its launch, the craft was still sending back data, even though its feeble signal measured about a billionth of a trillionth of a watt by the time it reached Earth, and hearing it required advanced processing techniques by NASA controllers. But with its power supply weakening, *Pioneer* could not maintain this tenuous link with its creators; by early 2003 its signals were no longer being received. Still, this trailblazer on the final frontier will always remind us what we can accomplish. As you read these words *Pioneer* continues onward, heading for the stars and bidding us to follow.

The first two components of the International Space Station (**opposite**) – the Russian built *Zarya* module (with solar panels), and the U.S. *Unity* module – seen from the shuttle *Atlantis* in May, 2000.

Lighting a dawn sky on March 8, 2001, the shuttle *Discovery* (**right**) ascends toward a rendezvous with the International Space Station at the start of the STS-102 mission. Aboard are Yuri Usachev, Jim Voss, and Susan Helms, the members of the Expedition Two crew, whose five-month tour of duty would include activation of the station's robotic arm.

Marsha Ivins, Ken Cockerel, and Mark Polansky peer at their spacewalking crewmates through an overhead window of the orbiter *Atlantis* during the STS-98 mission, February 2000. The astronauts installed the station's *Destiny* scientific laboratory.

Riding *Atlantis*' robotic arm, Jim Voss holds the main boom of the Russian-built *Strela* crane (**top**) before he and Jeff Williams install it on the station's *Zarya* module on May 21, 2000. This completed work begun on the crane a year earlier by spacewalkers Tammy Jernigan (**above right**) and Dan Barry. Bob Curbeam floats above the *Destiny* laboratory module on February 12, 2001 (**above**). Before the spacewalk, rookie Williams (**right**) gets a good-luck hug from Susan Helms.

Jim Voss of the Expedition Two crew consults an atlas (**left**) and makes repairs (**below**) in the Russian-built service module *Zvezda* during his five-month tour between March and August 2001. During the STS-98 mission Bob Curbeam (**below left**) and Ken Cockerel (**below**) install equipment racks in the U.S. *Destiny* laboratory.

Expedition One flight engineer Yuri Gidzenko (**above left**) floats within the *Leonardo* module, a "moving van" used to ferry supplies to and from the station. Expedition Two commander Yuri Usachev (**above**) shaves with an electric razor in the *Zvezda* module, March 28, 2001. Expedition One's Sergei Krikalev (**left**) views the approach the shuttle *Atlantis*, which is carrying the station's *Destiny* lab module, on February 9, 2001.

The International Space Station sails onward, March 18, 2001. The *Destiny* lab module is visible above the two giant solar panels that stretch 240 feet (73 metres), longer than the wingspan of a 747 jumbo jet. The crew of STS-102, which delivered the Expedition Two crew to the ISS, took this photo during a fly-around inspection of the station before heading back to Earth with the Expedition One crew.

Linda Godwin and Dan Tanni get a ride at the end of *Endeavour*'s robot arm on December 10, 2001. During a four-hour spacewalk the pair installed thermal blankets on the motors that rotate the space station's massive solar panels.

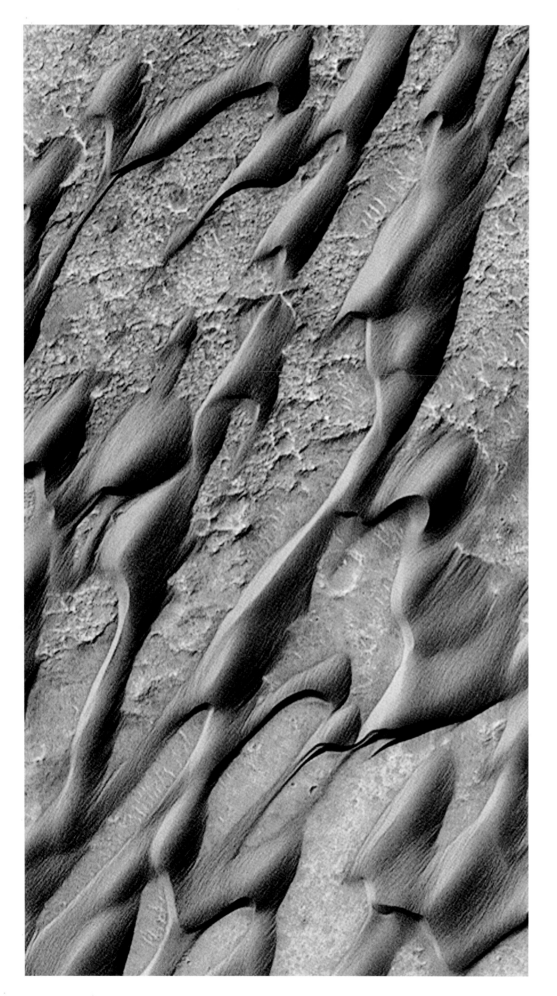

Scientists have a new and remarkably detailed view of Mars thanks to NASA's Mars *Global Surveyor* spacecraft, which has been circling the Red Planet since 1997. From its orbital height of more than 230 miles (370 kilometres), *Global Surveyor* can photograph objects the size of a compact car.

Wind is the dominant force shaping the Martian surface today. In this March 2001 view (**left**), which is about a mile (1.6 kilometres) wide, dunes of windblown dark sand cover the floor of an ancient crater in the planet's southern highlands. Grooves on the dunes' surfaces suggest that the dune sands have been cemented together and then eroded and scoured by wind.

A bizarre landscape on Mars' south polar ice cap (**opposite left**), resembling broken slices of Swiss cheese, was photographed in August, 1999. Scientists believe the circular depressions formed when frozen carbon dioxide, known on Earth as "dry ice," evaporated in the Martian sunlight. This view is 1.4 miles (2.25 kilometres) across.

Since 1972, when *Mariner 9* first discovered ancient Martian channels that had been carved by water, scientists have debated whether the channels were formed in sudden, catastrophic floods or by sustained, river-like flows. Analysis of this channel, 1.6 miles (2.6 kilometres) wide (**opposite right**) and located in Mars' northern plains, suggests that at least some of the Martian channels did form by sustained flow. *Global Surveyor* photographed it in January, 1998.

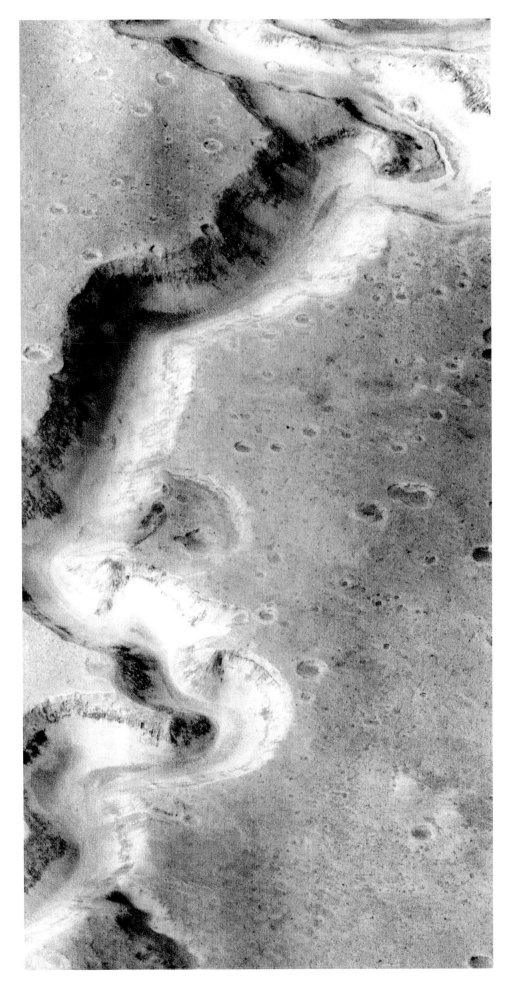

In June 2000 Mars *Global Surveyor* scientists announced that they had found signs that water has recently flowed – and may still flow – across the Martian surface. Among the evidence are these images, which show channels cut into the walls of a crater about 7 miles (11.3 kilometres) in diameter located in Mars' southern highlands. The channels are accompanied by debris deposits that strongly resemble drainage features on Earth, and may have been formed by running water on Mars; however, scientists are still debating their origin.

Orbiting 124 miles (200 kilometres) above the asteroid Eros, the NEAR-Shoemaker spacecraft snapped a mosaic of high-resolution images looking down on the asteroid's north pole. The large, shadowed crater on the upper surface is 3.3 miles (5.3 kilometres) across; Eros itself is 20 miles (32 kilometres) in length. NEAR-Shoemaker orbited Eros for a year beginning in February, 2000, in the first detailed reconnaissance of an asteroid.

As its final task at Eros, NEAR-Shoemaker made a descent to the asteroid's surface, touching down on February 12, 2001. Altitude 3,773 feet (1,150 metres), the image is about 180 feet (55 feet) across (**above**). Altitude 820 feet (250 metres); image is 39 feet (12 metres) across (**above right**). The final image from NEAR-Shoemaker (**right**) was made from an altitude of 394 feet (120 metres), showing an area 20 feet (6 metres) across. Part of a large boulder is visible at the top of the image; the bottom of the image was lost when transmission from the spacecraft was interrupted by touchdown on Eros.

The ragged, cratered nucleus of Comet Wild 2 is revealed in this close-up from NASA's *Stardust* spacecraft (**opposite**). *Stardust* made its close encounter with the icy wanderer on January 2, 2004 using an onboard collector to snare samples of the comet's dust particles for return to Earth. As a bonus, the craft sent back the most detailed images ever obtained of a cometary nucleus. *Stardust* and its precious cargo of comet dust will land on Earth in January 2006.

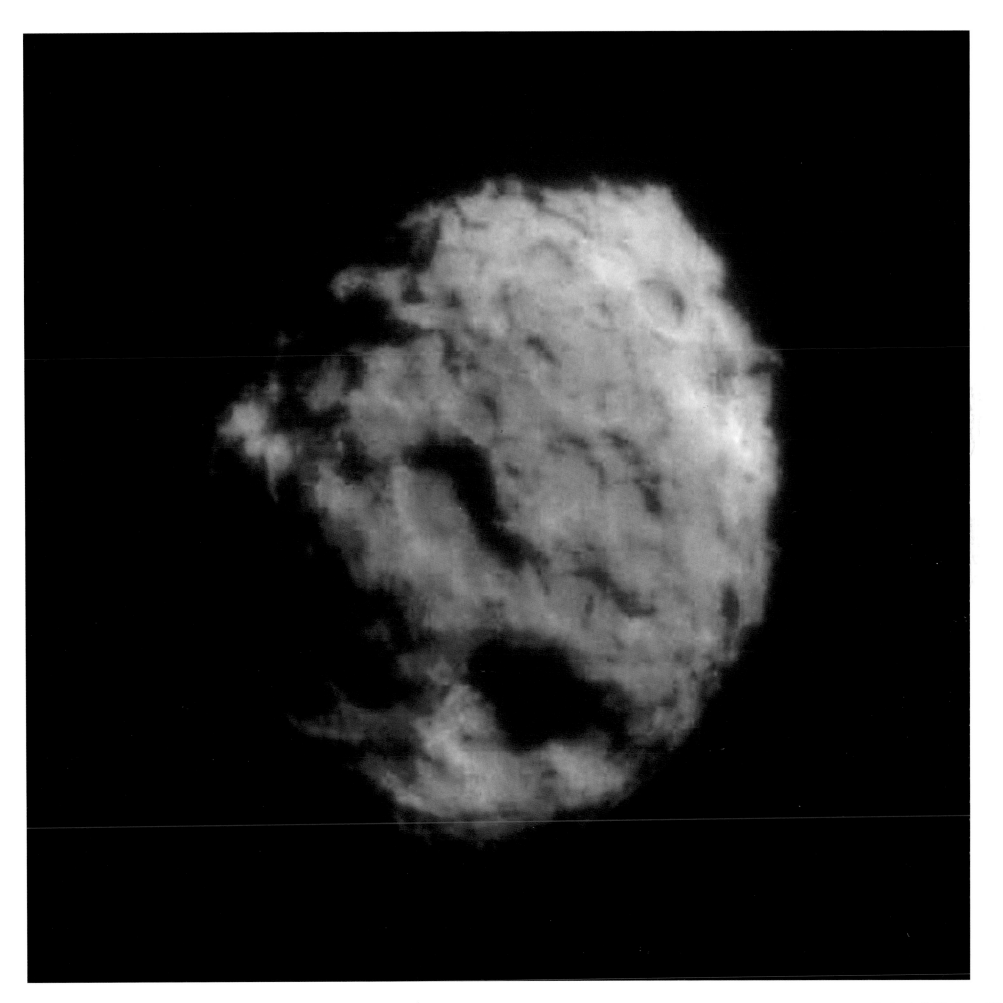

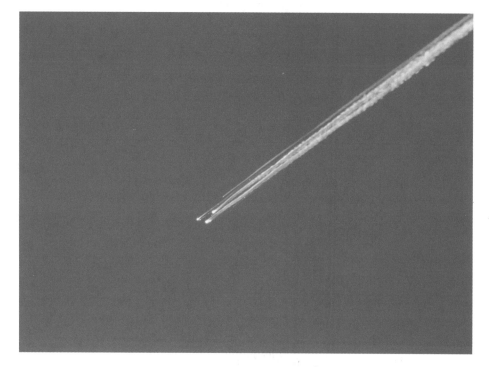

On February 1, 2003, after a 16-day science mission, the Space Shuttle *Columbia* disintegrated during reentry (**left**), killing its seven-member crew. Exposed film recovered from the debris included the astronauts' self-portrait (**above left**); clockwise from bottom: Rick Husband, Kalpana Chawla, David Brown, William McCool, Michael Anderson, Israeli astronaut Ilan Ramon, and Laurel Clark. A last-quarter moon seen above the Earth's horizon (**above**) was captured in an electronic photo taken by one of the crew and transmitted to Earth during the mission.

China's first manned spaceflight begins with the liftoff of the *Shenzhou 5* spacecraft (**left**) on October 15, 2003. Its lone occupant, 38-year-old pilot Yang Liwei (**below**), was photographed by an onboard television camera during the Earth-orbit mission, which lasted 21 hours, 23 minutes.

In January 2004 NASA's Mars Exploration Rovers, named *Spirit* and *Opportunity*, reached the martian surface to search for signs that water had shaped the Red Planet. Each 5-foot-long, 384-pound (1.5-metre, 174-kilogram) rover was equipped with high-resolution cameras, including a microscopic imager, and spectrometers to probe the chemical composition of the rocks and soil. On January 15, some 12 days after landing, *Spirit* looked back at its lander (**left**) after rolling its six wheels onto the rock-strewn floor of Gusev crater. On the other side of the planet, *Opportunity* sent back compelling evidence that the region called Meridiani Planum was once drenched in liquid water. Its microscopic imager revealed tiny spherules (**above**) that formed in a watery environment; this view, taken on February 12, 2004, is approximately 1 inch (2.5 centimetres) across.

Flagship of NASA's Earth Observing System, the Terra satellite has sent back stunning images since its launch in February 2000. **This page**: Australia's Great Barrier Reef on August 26, 2000.

Opposite: Hurricane Isabel about 400 miles (644 kilometres) north of Puerto Rico on September 14, 2003.

Icebergs break away from Antarctica's Ross Ice Shelf as seen by Terra on September 21, 2000.
Scientists monitor cracks in the Antarctic ice for clues to the potential effects of global warming.

ACKNOWLEDGEMENTS

The publishers would like to thank the following sources for their kind permission to reproduce the pictures in this book:

All photos supplied by NASA except for the following:

P1 NASA/Hamlyn Picture Library, 5 Art Dula, 15 Ronald Grant Archive, 16 Kaluga Museum, 17, 20 (both), 21 (left) Hulton Archive, 22 (top), 27 Novosti/Topham, 31 (right) National Air and Space Museum, 32 Novosti/Topham, 33 NASA/Genesis space Photo Library, 61 Center for Earth and Planetary Studies, 65 NASA/Hamlyn Picture Library, 75 left) Novosti/Topham, 86 NASA/Science Photo Library, 111 (top) NASA/Hamlyn Picture Library, 140 Moscow Space Research Institute, 141 (both) Energia Corporation,150 Andrew Chaikin collection,155 (top) NASA/Genesis Space Photo Library, 165 U.S. Geological Survey, 179 Hulton Archive, 188 Max Planck Institute, 191 Sky and Telescope magazine, 192, 193 (left) Art Dula, 193 (3 pics on right) Energia Corporation, 194 (left)Moscow Space Research Institute, 196, 197 Corbis/Roger Ressmeyer, 204-209 Space Telescope Science Institute, 214 (bottom) Novosti/Topham, 220 (top left) NASA/Science Photo Library, 222 U.S. Geological Survey, 226 Courtesy of SOHO/EIT Consortium. SOHO is a project of international cooperation between ESA and NASA, 235 left NASA/Science Photo Library, 246 (bottom) Corbis: Robert McCullough/Dallas Morning News, 247 Getty: AFP

22 (bottom), 23, 37 (left), 35, 36 (left), 37 (left), 37(right), 45, 54-5, 64 (top), 75 (right), 118-9, 119 (top), 141 (right), 151, 190 (left), 190 (top right), 190 (bottom right), 192 , 193 left), 193 (top right), 193 (centre right), 193 (bottom right), 194 (right), 195, 253 Private Collections of Ilona Velikaya, Moscow; Juriy Ivanchenko, Moscow; Svetlana Kasatkina, Kaluga; Space Museum, Moscow; National Space Museum, Kaluga.

Thanks to Amina Koltsova and Andrei Oldenburger for their help in scanning images from Russian collections.

Every effort has been made to acknowledge correctly and contact the source and/or copyright holder of each picture, and Carlton Books Limited apologises for any unintentional errors or omissions which will be corrected in future editions of this book.

Author's acknowledgements

Several people assisted in identifying and acquiring the spectacular images for this book. I am very grateful to Kipp Teague, creator of the website www.apolloarchive.com, for his excellent photographic research.

Thanks also to Mike Gentry and Gloria Sanchez at NASA's Johnson Space Center, Margaret Persinger at NASA's Kennedy Space Center, Rose Steinat at the National Air and Space Museum, Imelda Joson of Sky & Telescope magazine, J.L. Pickering, Art Dula, Tim Furniss and Vladimir Fishel.

I would also like to thank Sarah Larter and Clare Baggaley at Carlton Books for their hard work and for sharing my enthusiasm for this project.

Special thanks to my wife, Vicki, who shares my love of space exploration and is my best source of support, reality check and inspiration.

INDEX